ESSENTIALS OF GEOMETRY

Edited by
John Kristtofer H. San Pedro

ArclerPress

Essentials of Geometry

Edited by John Kristtofer H. San Pedro

Edition 2018

ArclerPress

Published by

Arcler Press LLC

708 3rd Avenue, 6th Floor

New York

NY 10017

United States of America

www.arclerpress.com

ISBN: 978-1-68094-584-3 (Hardback)

Printed in United States of America

Library of Congress Control Number: 2017933171

About the Editor

John Kristtofer H. San Pedro

John is a Mechanical Engineer and a Licensed Master Plumber from the Polytechnic University of the Philippines. With a successful history in creating innovative, cost efficient designs. A career focus on building engineering with experience in conceptual design through production. Highly skilled in drafting and using design software's such as AutoCAD, Revit, and Solidworks for his works. Been actively involved in Kaizen programs and achieve to be a champion in such competition during his 2 years stay in the automotive/manufacturing field.

Table of Contents

2 CONCEPTS AND FEATURES OF GEOMETRY

3 CLASSICAL BRANCHES

4 CONTEMPORARY BRANCHES OF GEOMETRY

5 ZERO / ONE-DIMENSIONAL GEOMETRY

6 TWO-DIMENSIONAL GEOMETRY

7 THREE-DIMENSIONAL GEOMETRY

8 FOUR-DIMENSIONAL SPACE GEOMETRY

List of Abbreviations

CGA	Conformal Geometric Algebra
CAD	Cylindrical Algebraic Decomposition
4d	Four-dimensional
1d	One-dimensional
STA	Spacetime algebra
3d	Three-dimensional
2d	Two-dimensional

Preface

Geometry is perhaps the most elementary of the sciences that enable man, by purely intellectual processes, to make predictions about physical world. The power of geometry, in the sense of accuracy and utility of these deductions, is impressive, and has been a powerful motivation for the study of logic in geometry. Geometry is most widely used application based mathematical concept used in our daily life. Geometry was evolved in 300 BC by Euclid known also as the father of Geometry, a Greek mathematician, was not meant only for mathematical concepts but for many normal applications for any person who has basic understanding of the concept. It is considered as one of the oldest concept and is concerned with shapes and sizes of the relative figures and with properties of space. Application of geometry in daily life consists of more than thousands of applications varying from lowest level to the highest end. Any physical space could be calculated by using the Geometrical concept. The entire Geometry is divided into two parts: one is plane Geometry and while the other is solid Geometry. Among the personalities, architects are one of the most extensive users of geometry. Defense applications are again one of the best examples of application of geometry in daily life. Daily operations of military consists much of real life Geometrical application such as loading of arms and piling of ammunition and firing of missiles. Locating the coordinates and then launching the missiles is another example of application of Geometry. At home based applications such as finding the square footage at the home for placing a refrigerator or locating a space to fit in an air conditioner requires calculation of the space required and this done by using area and volume concept. While whitewashing we need to make an estimate how much paint is required for the walls and roof, and without geometrical calculations it would be nearly impossible for us to calculate the amount of paint required. Another very interesting application of geometry in daily life is the use of it in computer graphics. Bio technology and Medical science also use geometric concepts too much for their daily operations. Large real estate developers use geometry in all their operations right from planning till deploying the plan. Modern geometry is thus a complete abstraction that crystallizes our ideas of the physical world, i.e., to start with.

Content Coverage

Chapter One presents the introduction of geometry as the power of geometry, in the sense of accuracy and utility of these deductions, is impressive, and has been a powerful motivation for the study of logic in geometry.

Chapter Two highlights the concepts and features of geometry. The concept of dimension is not restricted to physical objects. High-dimensional spaces frequently occur in mathematics and the sciences. They may be parameter spaces or configuration spaces such as in Lagrangian or Hamiltonian mechanics; these are abstract spaces, independent of the physical space we live in.

Chapter Three outlines the classical branches of geometry. Algebraic geometry occupies a central place in modern mathematics and has multiple conceptual connections with such diverse fields as complex analysis, topology and number theory. Initially a study of systems of polynomial equations in several variables, the subject.

Chapter Four highlights about contemporary branches of geometry. In absolute geometry it is also provable that two lines perpendicular to the same line cannot intersect (which makes the two lines parallel by definition of parallel lines), proving that the summit angles of an Saccheri quadrilateral cannot be obtuse , and that spherical geometry is not an absolute geometry.

Chapter Five focuses on one dimensional geometry. The real number line and the real projective line are one-dimensional, though the latter is homeomorphic to a circle. In algebraic geometry there are several structures which are technically one-dimensional spaces but referred to in other terms.

Chapter Six emphasizes on two-dimensional space or bi-dimensional space is a geometric model of the planar projection of the physical universe. The two dimensions are commonly called length and width. Both directions lie in the same plane.

Chapter Seven focuses on three-dimensional geometric setting in which three values (called parameters) are required to determine the position of an element (i.e., point).

Chapter Eight presents a framework on four-dimensional space. It typically is more specifically four-dimensional Euclidean space, generalizing the rules of three-dimensional Euclidean space.

CHAPTER 1

Geometry: History and Scope

Chapter Contents

Geometry is perhaps the most elementary of the sciences that enable man, by purely intellectual processes, to make predictions (based on observation) about physical world. The power of geometry, in the sense of accuracy and utility of these deductions, is impressive, and has been a powerful motivation for the study of logic in geometry.

INTRODUCTION

Geometry is a branch of mathematics concerned with questions of shape, size, relative position of figures, and the properties of space. A mathematician who works in the field of geometry is called a geometer.

Geometry arose independently in a number of early cultures as a practical way for dealing with lengths, areas, and volumes. Geometry began to see elements of formal mathematical science emerging in the West as early as the 6th century BC. By the 3rd century BC, geometry was put into an axiomatic form by Euclid, whose treatment, Euclid›s Elements, set a standard for many centuries to follow. Geometry arose independently in

India, with texts providing rules for geometric constructions appearing as early as the 3rd century BC. Islamic scientists preserved Greek ideas and expanded on them during the Middle Ages. By the early 17th century, geometry had been put on a solid analytic footing by mathematicians such as René Descartes and Pierre de Fermat. Since then, and into modern times, geometry has expanded into non-Euclidean geometry and manifolds, describing spaces that lie beyond the normal range of human experience.

While geometry has evolved significantly throughout the years, there are some general concepts that are more or less fundamental to geometry. These include the concepts of points, lines, planes, surfaces, angles, and curves, as well as the more advanced notions of manifolds and topology or metric.

Contemporary geometry has many subfields:

- **Euclidean geometry** is geometry in its classical sense. The mandatory educational curriculum of the majority of nations includes the study of points, lines, planes, angles, triangles, congruence, similarity, solid figures, circles, and analytic geometry. Euclidean geometry also has applications in computer science, crystallography, and various branches of modern mathematics.
- **Differential geometry** uses techniques of calculus and linear algebra to study problems in geometry. It has applications in physics, including in general relativity.
- **Topology** is the field concerned with the properties of geometric objects that are unchanged by continuous mappings. In practice, this often means dealing with large-scale properties of spaces, such as connectedness and compactness.
- **Convex geometry** investigates convex shapes in the Euclidean space and its more abstract analogues, often using techniques of real analysis. It has close connections to convex analysis, optimization and functional analysis and important applications in number theory.
- **Algebraic geometry** studies geometry through the use of multivariate polynomials and other algebraic techniques. It has applications in many areas, including cryptography and string theory.
- **Discrete geometry** is concerned mainly with questions of relative position of simple geometric objects, such as points, lines and circles. It shares many methods and principles with combinatorics.

Geometry has applications to many fields, including art, architecture, physics, as well as to other branches of mathematics.

HISTORY OF GEOMETRY

Geometry (from the Ancient Greek: γεωμετρία; geo- "earth", -metron "measurement") arose as the field of knowledge dealing with spatial relationships. Geometry was one of the two fields of pre-modern mathematics, the other being the study of numbers (arithmetic).

Classic geometry was focused in compass and straightedge constructions. Geometry was revolutionized by Euclid, who introduced mathematical rigor and the axiomatic method still in use today. His book, The Elements is widely considered the most influential textbook of all time, and was known to all educated people in the West until the middle of the

20th century. In modern times, geometric concepts have been generalized to a high level of abstraction and complexity, and have been subjected to the methods of calculus and abstract algebra, so that many modern branches of the field are barely recognizable as the descendants of early geometry. (See Areas of mathematics and Algebraic geometry.)

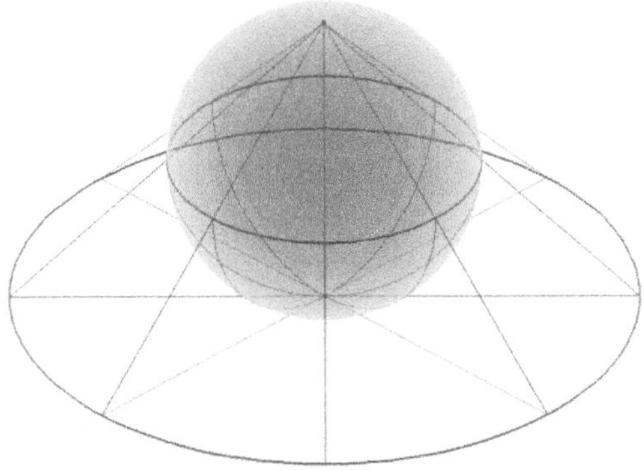

Projecting a sphere to a plane.

Early Geometry

The earliest recorded beginnings of geometry can be traced to early peoples, who discovered obtuse triangles in the ancient Indus Valley (see Harappan Mathematics), and ancient Babylonia (see Babylonian mathematics) from around 3000 BC. Early geometry was a collection of empirically discovered principles concerning lengths, angles, areas, and volumes, which were developed to meet some practical need in surveying, construction, astronomy, and various crafts. Among these were some surprisingly sophisticated principles, and a modern mathematician might be hard put to derive some of them without the use of calculus. For example, both the Egyptians and the Babylonians were aware of versions of the Pythagorean theorem about 1500 years before Pythagoras and the Indian Sulba Sutras around 800 B.C. contained the first statements of the theorem; the Egyptians had a correct formula for the volume of a frustum of a square pyramid;

Egyptian Geometry

The ancient Egyptians knew that they could approximate the area of a circle as follows:

Area of Circle $\approx$ [(Diameter) x 8/9]^2.

Problem 30 of the Ahmes papyrus uses these methods to calculate the area of a circle, according to a rule that the area is equal to the square of 8/9 of the circle's diameter. This assumes that π is $4\times(8/9)^2$ (or 3.160493...), with an error of slightly over 0.63 percent. This value was slightly less accurate than the calculations of the Babylonians (25/8 = 3.125, within 0.53 percent), but was not otherwise surpassed until Archimedes› approximation of 211875/67441 = 3.14163, which had an error of just over 1 in 10,000.

Interestingly, Ahmes knew of the modern 22/7 as an approximation for π, and used it to split a hekat, hekat x 22/x x 7/22 = hekat; however, Ahmes continued to use the traditional 256/81 value for π for computing his hekat volume found in a cylinder.

Problem 48 involved using a square with side 9 units. This square was cut into a 3x3 grid. The diagonal of the corner squares were used to make an irregular octagon with an area of 63 units. This gave a second value for π of 3.111...

The two problems together indicate a range of values for π between 3.11 and 3.16.

Problem 14 in the Moscow Mathematical Papyrus gives the only ancient example finding the volume of a frustum of a pyramid, describing the correct formula:

$$V = \frac{1}{3}h(x_1^2 + x_1x_2 + x_2^2).$$

Babylonian Geometry

The Babylonians may have known the general rules for measuring areas and volumes. They measured the circumference of a circle as three times the diameter and the area as one-twelfth the square of the circumference, which would be correct if π is estimated as 3. The volume of a cylinder was taken as the product of the base and the height, however, the volume of the frustum of a cone or a square pyramid was incorrectly taken as the product of the height and half the sum of the bases. The Pythagorean theorem was also known to the Babylonians. Also, there was a recent discovery in which a tablet used π as 3 and 1/8. The Babylonians are also known for the Babylonian mile, which was a measure of distance equal to about seven miles today. This measurement for distances eventually was converted to a time-mile used for measuring the travel of the Sun, therefore, representing time. There have been recent discoveries showing that ancient Babylonians may have discovered astronomical geometry nearly 1400 years before Europeans did.

Vedic India

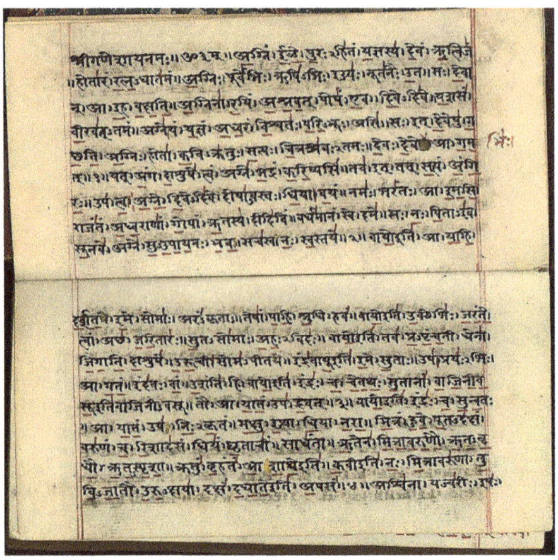

Rigveda manuscript in Devanagari.

The Indian Vedic period had a tradition of geometry, mostly expressed in the construction of elaborate altars. Early Indian texts (1st millennium BC) on this topic include the Satapatha Brahmana and the Śulba Sūtras.

According to (Hayashi 2005, p. 363), the Śulba Sūtras contain "the earliest extant verbal expression of the Pythagorean Theorem in the world, although it had already been known to the Old Babylonians."

The diagonal rope of an oblong (rectangle) produces both which the flank (pārśvamāni) and the horizontal <ropes> produce separately."

They contain lists of Pythagorean triples, which are particular cases of Diophantine equations. They also contain statements (that with hindsight we know to be approximate) about squaring the circle and "circling the square."

The Baudhayana Sulba Sutra, the best-known and oldest of the Sulba Sutras (dated to the 8th or 7th century BC) contains examples of simple Pythagorean triples, such as: $(3,4,5)$, $(5,12,13)$, $(8,15,17)$, $(7,24,25)$, and $(12,35,37)$ as well as a statement of the Pythagorean theorem for the sides of a square: "The rope which is stretched across the diagonal of a square produces an area double the size of the original square." It also contains the general statement of the Pythagorean theorem (for the sides of a rectangle): "The rope stretched along the length of the diagonal of a rectangle makes an area which the vertical and horizontal sides make together."

According to mathematician S. G. Dani, the Babylonian cuneiform tablet Plimpton 322 written c. 1850 BC "contains fifteen Pythagorean triples with quite large entries, including (13500, 12709, 18541) which is a primitive triple, indicating, in particular, that there was sophisticated understanding on the topic" in Mesopotamia in 1850 BC. "Since these tablets predate the Sulbasutras period by several centuries, taking into account the contextual appearance of some of the triples, it is reasonable to expect that similar understanding would have been there in India."Dani goes on to say:

"As the main objective of the Sulvasutras was to describe the constructions of altars and the geometric principles involved in them, the subject of Pythagorean triples, even if it had been well understood may still not have featured in the Sulvasutras. The occurrence of the triples in the Sulvasutras is comparable to mathematics that one may encounter in an introductory book on architecture or another similar applied area, and would not correspond directly to the overall knowledge on the topic at that time. Since, unfortunately, no other contemporaneous sources have been found it may never be possible to settle this issue satisfactorily."

In all, three Sulba Sutras were composed. The remaining two, the Manava Sulba Sutra composed by Manava (fl. 750-650 BC) and the Apastamba Sulba Sutra, composed by Apastamba (c. 600 BC), contained results similar to the Baudhayana Sulba Sutra.

Greek Geometry

Classical Greek Geometry

For the ancient Greek mathematicians, geometry was the crown jewel of their sciences, reaching a completeness and perfection of methodology that no other branch of their knowledge had attained. They expanded the range of geometry to many new kinds of figures, curves, surfaces, and solids; they changed its methodology from trial-and-error to logical deduction; they recognized that geometry studies "eternal forms", or abstractions, of which physical objects are only approximations; and they developed the idea of the "axiomatic method", still in use today.

Thales and Pythagoras

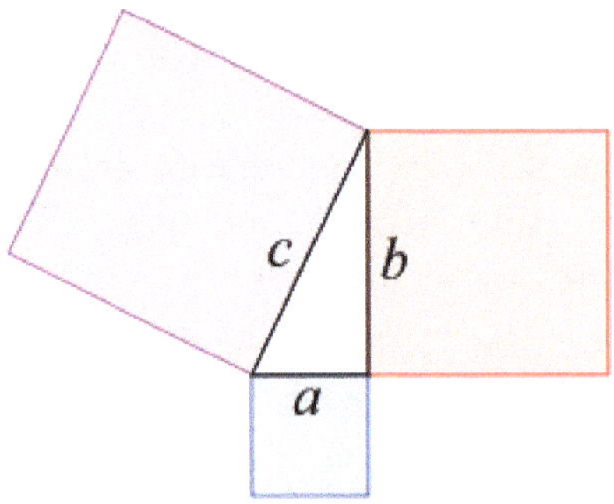

Pythagorean theorem: $a^2 + b^2 = c^2$

Thales (635-543 BC) of Miletus (now in southwestern Turkey), was the first to whom deduction in mathematics is attributed. There are five geometric propositions for which he wrote deductive proofs, though his proofs have not survived. Pythagoras (582-496 BC) of Ionia, and later, Italy, then colonized by Greeks, may have been a student of Thales, and traveled to Babylon and Egypt. The theorem that bears his name may not have been his discovery, but he was probably one of the first to give a deductive proof of it. He gathered a group of students around him to study mathematics, music, and philosophy, and together they discovered most of what high school students learn today in their geometry courses. In addition, they made the profound discovery of incommensurable lengths and irrational numbers.

Plato

Plato (427-347 BC) is a philosopher that is highly esteemed by the Greeks. There is a story that he had inscribed above the entrance to his famous school, "Let none ignorant of geometry enter here." However, the story is considered to be untrue. Though he was not a mathematician himself, his views on mathematics had great influence. Mathematicians thus accepted his belief that geometry should use no tools but compass and straightedge – never measuring instruments such as a marked ruler or a protractor, because these were a workman's tools, not worthy of a scholar. This dictum led to a deep study of possible compass and straightedge constructions, and three classic construction problems: how to use these tools to trisect an angle, to construct a cube twice the volume of a given cube, and to construct a square equal in area to a given circle. The proofs of the impossibility of these constructions, finally achieved in the 19th century, led to important principles regarding the deep structure of the real number system. Aristotle (384-322 BC), Plato's greatest pupil, wrote a treatise on methods of reasoning used in deductive proofs (see Logic) which was not substantially improved upon until the 19th century.

Hellenistic Geometry

Euclid

Statue of Euclid in the Oxford University Museum of Natural History.

Woman teaching geometry. Illustration at the beginning of a medieval translation of Euclid's Elements, (c. 1310)

Euclid (c. 325-265 BC), of Alexandria, probably a student at the Academy founded by Plato, wrote a treatise in 13 books (chapters), titled The Elements of Geometry, in which he presented geometry in an ideal axiomatic form, which came to be known as Euclidean geometry. The treatise is not a compendium of all that the Hellenistic mathematicians knew at the time about geometry; Euclid himself wrote eight more advanced books on geometry. We know from other references that Euclid's was not the first elementary geometry textbook, but it was so much superior that the others fell into disuse and were lost. He was brought to the university at Alexandria by Ptolemy I, King of Egypt.

The Elements began with definitions of terms, fundamental geometric principles (called axioms or postulates), and general quantitative principles (called common notions) from which all the rest of geometry could be logically deduced. Following are his five axioms, somewhat paraphrased to make the English easier to read.

- Any two points can be joined by a straight line.
- Any finite straight line can be extended in a straight line.
- A circle can be drawn with any center and any radius.
- All right angles are equal to each other.
- If two straight lines in a plane are crossed by another straight line (called the transversal), and the interior angles between the two lines and the transversal

lying on one side of the transversal add up to less than two right angles, then on that side of the transversal, the two lines extended will intersect (also called the parallel postulate).

Archimedes

Archimedes (287-212 BC), of Syracuse, Sicily, when it was a Greek city-state, is often considered to be the greatest of the Greek mathematicians, and occasionally even named as one of the three greatest of all time (along with Isaac Newton and Carl Friedrich Gauss). Had he not been a mathematician, he would still be remembered as a great physicist, engineer, and inventor. In his mathematics, he developed methods very similar to the coordinate systems of analytic geometry, and the limiting process of integral calculus. The only element lacking for the creation of these fields was an efficient algebraic notation in which to express his concepts.

After Archimedes

Geometry was connected to the divine for most medieval scholars. The compass in this 13th-century manuscript is a symbol of God's act of Creation.

After Archimedes, Hellenistic mathematics began to decline. There were a few minor stars yet to come, but the golden age of geometry was over. Proclus (410-485), author of Commentary on the First Book of Euclid, was one of the last important players in Hellenistic geometry. He was a competent geometer, but more importantly, he was a superb commentator on the works that preceded him. Much of that work did not survive to modern times, and is known to us only through his commentary. The Roman Republic and Empire that succeeded and absorbed the Greek city-states produced excellent engineers, but no mathematicians of note.

The great Library of Alexandria was later burned. There is a growing consensus among historians that the Library of Alexandria likely suffered from several destructive events, but that the destruction of Alexandria's pagan temples in the late 4th century was probably the most severe and final one. The evidence for that destruction is the most definitive and secure. Caesar's invasion may well have led to the loss of some 40,000-70,000 scrolls in a warehouse adjacent to the port (as Luciano Canfora argues, they were likely copies produced by the Library intended for export), but it is unlikely to have affected the Library or Museum, given that there is ample evidence that both existed later.

Civil wars, decreasing investments in maintenance and acquisition of new scrolls and generally declining interest in non-religious pursuits likely contributed to a reduction in the body of material available in the Library, especially in the 4th century. The Serapeum was certainly destroyed by Theophilus in 391, and the Museum and Library may have fallen victim to the same campaign.

Classical Indian Geometry

In the Bakhshali manuscript, there is a handful of geometric problems (including problems about volumes of irregular solids). The Bakhshali manuscript also "employs a decimal place value system with a dot for zero." Aryabhata's Aryabhatiya (499) includes the computation of areas and volumes.

Brahmagupta wrote his astronomical work Brāhma Sphua Siddhānta in 628. Chapter 12, containing 66 Sanskrit verses, was divided into two sections: "basic operations" (including cube roots, fractions, ratio and proportion, and barter) and "practical mathematics" (including mixture, mathematical series, plane figures, stacking bricks, sawing of timber, and piling of grain). In the latter section, he stated his famous theorem on the diagonals of a cyclic quadrilateral:

Brahmagupta›s theorem: If a cyclic quadrilateral has diagonals that are perpendicular to each other, then the perpendicular line drawn from the point of intersection of the diagonals to any side of the quadrilateral always bisects the opposite side.

Chapter 12 also included a formula for the area of a cyclic quadrilateral (a generalization of Heron›s formula), as well as a complete description of rational triangles (i.e. triangles with rational sides and rational areas).

Brahmagupta›s formula: The area, A, of a cyclic quadrilateral with sides of lengths a, b, c, d, respectively, is given by

$$A = \sqrt{(s-a)(s-b)(s-c)(s-d)}$$

where s, the semiperimeter, given by: $s = \frac{a+b+c+d}{2}.$

Brahmagupta's Theorem on rational triangles: A triangle with rational sides a, b, c and rational area is of the form:

$$a = \frac{u^2}{v} + v, \quad b = \frac{u^2}{w} + w, \quad c = \frac{u^2}{v} + \frac{u^2}{w} - (v + w)$$

for some rational numbers $u, v,$ and w.

Chinese Geometry

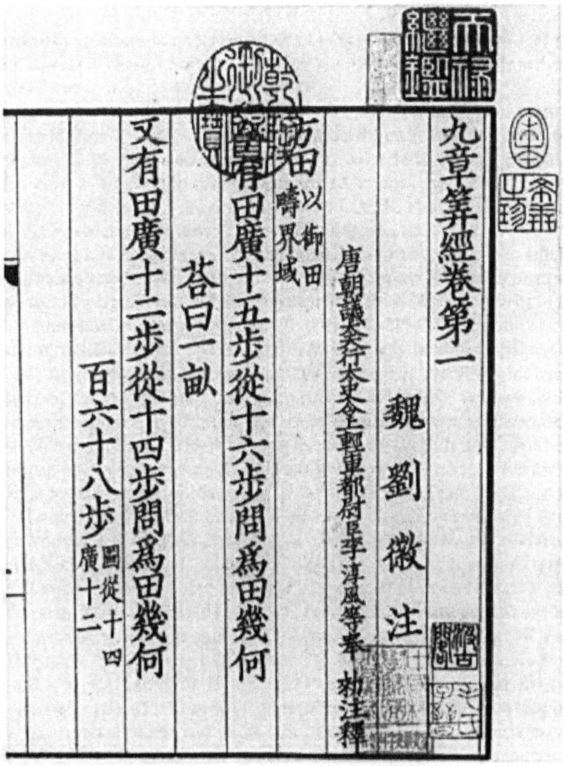
九章筭經卷第一 魏 劉徽 注
唐朝議大夫行太史令上輕車都尉臣李淳風等奉 勑注釋
方田 以御田疇界域
今有田廣十五步從十六步問爲田幾何
荅曰一畝
又有田廣十二步從十四步問爲田幾何
百六十八步 圖從十四廣十二

The Nine Chapters on the Mathematical Art, first compiled in 179 AD, with added commentary in the 3rd century by Liu Hui.

Haidao Suanjing, Liu Hui, 3rd century.

The first definitive work (or at least oldest existent) on geometry in China was the Mo Jing, the Mohist canon of the early philosopher Mozi (470-390 BC). It was compiled years after his death by his followers around the year 330 BC. Although the Mo Jing is the oldest existent book on geometry in China, there is the possibility that even older written material existed. However, due to the infamous Burning of the Books in a political maneuver by the Qin Dynasty ruler Qin Shihuang (r. 221-210 BC), multitudes of written literature created before his time were purged. In addition, the Mo Jing presents geometrical concepts in mathematics that are perhaps too advanced not to have had a previous geometrical base or mathematic background to work upon.

The Mo Jing described various aspects of many fields associated with physical science, and provided a small wealth of information on mathematics as well. It provided an 'atomic' definition of the geometric point, stating that a line is separated into parts, and the part which has no remaining parts (i.e. cannot be divided into smaller parts) and thus forms the extreme end of a line is a point. Much like Euclid's first and third definitions and Plato's 'beginning of a line', the Mo Jing stated that "a point may stand at the end (of a line) or at its beginning like a head-presentation in childbirth. (As to its invisibility) there is nothing similar to it.Similar to the atomists of Democritus, the Mo Jing stated that a point is the smallest unit, and cannot be cut in half, since 'nothing' cannot be halved. It stated that two lines of equal length will always finish at the same place, while providing definitions for the comparison of lengths and for parallels, along with principles of space and bounded space. It also described the fact that planes without the quality of thickness cannot be piled up since they cannot mutually touch. The book provided definitions for circumference, diameter, and radius, along with the definition of volume.

The Han Dynasty (202 BC-220 AD) period of China witnessed a new flourishing of mathematics. One of the oldest Chinese mathematical texts to present geometric progressions was the Suàn shù shū of 186 BC, during the Western Han era. The mathematician, inventor, and astronomer Zhang Heng (78-139 AD) used geometrical formulas to solve mathematical problems. Although rough estimates for pi (π) were given in the Zhou Li (compiled in the 2nd century BC), it was Zhang Heng who was the first to make a concerted effort at creating a more accurate formula for pi. Zhang Heng approximated pi as 730/232 (or approx 3.1466), although he used another formula of pi in finding a spherical volume, using the square root of 10 (or approx 3.162) instead. Zu Chongzhi (429-500 AD) improved the accuracy of the approximation of pi to between 3.1415926 and 3.1415927, with $^{355}/_{113}$ (密率, Milü, detailed approximation) and $^{22}/_{7}$ (约率, Yuelü, rough approximation) being the other notable approximation. In comparison to later works, the formula for pi given by the French mathematician Franciscus Vieta (1540-1603) fell halfway between Zu›s approximations.

The Nine Chapters on the Mathematical Art

The Nine Chapters on the Mathematical Art, the title of which first appeared by 179 AD on a bronze inscription, was edited and commented on by the 3rd century mathematician Liu Hui from the Kingdom of Cao Wei. This book included many problems where geometry was applied, such as finding surface areas for squares and circles, the volumes of solids in various three-dimensional shapes, and included the use of the Pythagorean theorem. The book provided illustrated proof for the Pythagorean theorem, contained a written dialogue between of the earlier Duke of Zhou and Shang Gao on the properties of the right angle triangle and the Pythagorean theorem, while also referring to the astronomical gnomon, the circle and square, as well as measurements of heights and distances. The editor Liu Hui listed pi as 3.141014 by using a 192 sided polygon, and then calculated pi as 3.14159 using a 3072 sided polygon. This was more accurate than Liu Hui's contemporary Wang Fan, a mathematician and astronomer from Eastern Wu, would render pi as 3.1555 by using $^{142}/_{45}$. Liu Hui also wrote of mathematical surveying to calculate distance measurements of depth, height, width, and surface area. In terms of solid geometry, he figured out that a wedge with rectangular base and both sides sloping could be broken down into a pyramid and a tetrahedral wedge. He also figured out that a wedge with trapezoid base and both sides sloping could be made to give two tetrahedral wedges separated by a pyramid. Furthermore, Liu Hui described Cavalieri's principle on volume, as well as Gaussian elimination. From the Nine Chapters, it listed the following geometrical formulas that were known by the time of the Former Han Dynasty (202 BCE–9 CE).

Areas for the

• Square • Rectangle • Circle • Isosceles triangle	• Rhomboid • Trapezoid • Double trapezium • Segment of a circle • Annulus ('ring' between two concentric circles)		

Volumes for the

• Parallelepiped with two square surfaces • Parallelepiped with no square surfaces • Pyramid • Frustum of pyramid with square base • Frustum of pyramid with rectangular base of unequal sides	• Cube • Prism • Wedge with rectangular base and both sides sloping • Wedge with trapezoid base and both sides sloping • Tetrahedral wedge	• Frustum of a wedge of the second type (used for applications in engineering) • Cylinder • Cone with circular base • Frustum of a cone • Sphere	

Continuing the geometrical legacy of ancient China, there were many later figures to come, including the famed astronomer and mathematician Shen Kuo (1031-1095 CE), Yang Hui (1238-1298) who discovered Pascal's Triangle, Xu Guangqi (1562-1633), and many others.

Islamic Golden Age

The final destruction of the Library of Alexandria at the Muslim conquest of Egypt in AD 642 marks the collapse of classical antiquity in the West, and the beginning of the European "Dark Ages". By the beginning of the 9th century, the "Islamic Golden Age" flourished, the establishment of the House of Wisdom in Baghdad marking a separate tradition of science in the medieval Islamic world, building not only Hellenistic but also on Indian sources.

Page from the Al-Jabr wa-al-Muqabilah

Although the Islamic mathematicians are most famed for their work on algebra, number theory and number systems, they also made considerable contributions to geometry, trigonometry and mathematical astronomy, and were responsible for the development of algebraic geometry.

Al-Mahani (born 820) conceived the idea of reducing geometrical problems such as duplicating the cube to problems in algebra. Al-Karaji (born 953) completely freed algebra from geometrical operations and replaced them with the arithmetical type of operations which are at the core of algebra today.

Thābit ibn Qurra (known as Thebit in Latin) (born 836) contributed to a number of areas in mathematics, where he played an important role in preparing the way for such important mathematical discoveries as the extension of the concept of number to (positive) real numbers, integral calculus, theorems in spherical trigonometry, analytic geometry, and non-Euclidean geometry. In astronomy Thabit was one of the first reformers of the Ptolemaic system, and in mechanics he was a founder of statics. An important geometrical aspect of Thabit›s work was his book on the composition of ratios. In this book, Thabit deals with arithmetical operations applied to ratios of geometrical quantities. The Greeks had dealt with geometric quantities but had not thought of them in the same way as numbers to which the usual rules of arithmetic could be applied. By introducing arithmetical operations on quantities previously regarded as geometric and non-numerical, Thabit started a trend which led eventually to the generalisation of the number concept.

In some respects, Thabit is critical of the ideas of Plato and Aristotle, particularly regarding motion. It would seem that here his ideas are based on an acceptance of using arguments concerning motion in his geometrical arguments. Another important contribution Thabit made to geometry was his generalization of the Pythagorean theorem, which he extended from special right triangles to all triangles in general, along with a general proof.

Ibrahim ibn Sinan ibn Thabit (born 908), who introduced a method of integration more general than that of Archimedes, and al-Quhi (born 940) were leading figures in a revival and continuation of Greek higher geometry in the Islamic world. These mathematicians, and in particular Ibn al-Haytham, studied optics and investigated the optical properties of mirrors made from conic sections.

Astronomy, time-keeping and geography provided other motivations for geometrical and trigonometrical research. For example Ibrahim ibn Sinan and his grandfather Thabit ibn Qurra both studied curves required in the construction of sundials. Abu›l-Wafa and Abu Nasr Mansur both applied spherical geometry to astronomy.

A 2007 paper in the journal Science suggested that girih tiles possessed properties consistent with self-similar fractal quasicrystalline tilings such as the Penrose tilings.

Renaissance

An engraving by Albrecht Dürer featuring Mashallah, from the title page of the De scientia motus orbis (Latin version with engraving, 1504). As in many medieval illustrations, the compass here is an icon of religion as well as science, in reference to God as the architect of creation

The transmission of the Greek Classics to medieval Europe via the Arabic literature of the 9th to 10th century Islamic Golden Age began in the 10th century and culminated in the Latin translations of the 12th century. A copy of Ptolemy›s Almagest was brought back to Sicily by Henry Aristippus (d. 1162), as a gift from the Emperor to King William I (r. 1154–1166). An anonymous student at Salerno travelled to Sicily and translated the Almagest as well as several works by Euclid from Greek to Latin. Although the Sicilians generally translated directly from the Greek, when Greek texts were not available, they would translate from Arabic. Eugenius of Palermo (d. 1202) translated Ptolemy's Optics into Latin, drawing on his knowledge of all three languages in the task. The rigorous deductive methods of geometry found in Euclid's Elements of Geometry were relearned, and further development of geometry in the styles of both Euclid (Euclidean geometry) and Khayyam (algebraic geometry) continued, resulting in an abundance of new theorems and concepts, many of them very profound and elegant.

Advances in the treatment of perspective were made in Renaissance art of the 14th to 15th century which went beyond what had been achieved in antiquity. In Renaissance architecture of the Quattrocento, concepts of architectural order were explored and rules were formulated. A prime example of is the Basilica di San Lorenzo in Florence by Filippo Brunelleschi (1377–1446).

In c. 1413 Filippo Brunelleschi demonstrated the geometrical method of perspective, used today by artists, by painting the outlines of various Florentine buildings onto a mirror. Soon after, nearly every artist in Florence and in Italy used geometrical perspective in their paintings, notably Masolino da Panicale and Donatello. Melozzo da Forlì first used the technique of upward foreshortening (in Rome, Loreto, Forlì and others), and was celebrated for that. Not only was perspective a way of showing depth, it was also a new method of composing a painting. Paintings began to show a single, unified scene, rather than a combination of several.

As shown by the quick proliferation of accurate perspective paintings in Florence, Brunelleschi likely understood (with help from his friend the mathematician Toscanelli), but did not publish, the mathematics behind perspective. Decades later, his friend Leon Battista Alberti wrote De pictura (1435/1436), a treatise on proper methods of showing distance in painting based on Euclidean geometry. Alberti was also trained in the science of optics through the school of Padua and under the influence of Biagio Pelacani da Parma who studied Alhazen›s Optics'.

Piero della Francesca elaborated on Della Pittura in his De Prospectiva Pingendi in the 1470s. Alberti had limited himself to figures on the ground plane and giving an overall basis for perspective. Della Francesca fleshed it out, explicitly covering solids in any area of the picture plane. Della Francesca also started the now common practice of using illustrated figures to explain the mathematical concepts, making his treatise easier to understand than Alberti›s. Della Francesca was also the first to accurately draw the Platonic solids as they would appear in perspective.

Perspective remained, for a while, the domain of Florence. Jan van Eyck, among others, was unable to create a consistent structure for the converging lines in paintings, as in London›s The Arnolfini Portrait, because he was unaware of the theoretical breakthrough just then occurring in Italy. However he achieved very subtle effects by manipulations of scale in his interiors. Gradually, and partly through the movement of academies of the arts, the Italian techniques became part of the training of artists across Europe, and later other parts of the world. The culmination of these Renaissance traditions finds its ultimate synthesis in the research of the architect, geometer, and optician Girard Desargues on perspective, optics and projective geometry.

The Vitruvian Man by Leonardo da Vinci(c. 1490) depicts a man in two superimposed positions with his arms and legs apart and inscribed in a circle and square. The drawing is based on the correlations of ideal human proportions with geometry described by the ancient Roman architect Vitruvius in Book III of his treatise De Architectura.

Modern Geometry

The 17th Century

DISCOURS
DE LA METHODE
Pour bien conduire sa raison, & chercher
la verité dans les sciences.
PLUS
LA DIOPTRIQVE.
LES METEORES.
ET
LA GEOMETRIE.
Qui sont des essais de cete METHODE.

A LEYDE
De l'Imprimerie de IAN MAIRE.
CIↃ IↃ C XXXVII.
Avec Privilege.

Discourse on Method by René Descartes

In the early 17th century, there were two important developments in geometry. The first and most important was the creation of analytic geometry, or geometry with coordinates and equations, by René Descartes (1596–1650) and Pierre de Fermat (1601–1665). This was a necessary precursor to the development of calculus and a precise quantitative science of physics. The second geometric development of this period was the systematic study of projective geometry by Girard Desargues (1591–1661). Projective geometry is the study of geometry without measurement, just the study of how points align with each other. There had been some early work in this area by Hellenistic geometers, notably Pappus (c. 340). The greatest flowering of the field occurred with Jean-Victor Poncelet (1788–1867).

In the late 17th century, calculus was developed independently and almost simultaneously by Isaac Newton (1642–1727) and Gottfried Wilhelm Leibniz (1646–1716). This was the beginning of a new field of mathematics now called analysis. Though not itself a branch of geometry, it is applicable to geometry, and it solved two families of problems that had long been almost intractable: finding tangent lines to odd curves, and finding areas enclosed by those curves. The methods of calculus reduced these problems mostly to straightforward matters of computation.

The 18th and 19th Centuries

Non-Euclidean Geometry

The very old problem of proving Euclid's Fifth Postulate, the "Parallel Postulate", from his first four postulates had never been forgotten. Beginning not long after Euclid, many attempted demonstrations were given, but all were later found to be faulty, through allowing into the reasoning some principle which itself had not been proved from the first four postulates. Though Omar Khayyám was also unsuccessful in proving the parallel postulate, his criticisms of Euclid's theories of parallels and his proof of properties of figures in non-Euclidean geometries contributed to the eventual development of non-Euclidean geometry. By 1700 a great deal had been discovered about what can be proved from the first four, and what the pitfalls were in attempting to prove the fifth. Saccheri, Lambert, and Legendre each did excellent work on the problem in the 18th century, but still fell short of success. In the early 19th century, Gauss, Johann Bolyai, and Lobatchewsky, each independently, took a different approach. Beginning to suspect that it was impossible to prove the Parallel Postulate, they set out to develop a self-consistent geometry in which that postulate was false. In this they were successful, thus creating the first non-Euclidean geometry. By 1854, Bernhard Riemann, a student of Gauss, had applied methods of calculus in a ground-breaking study of the intrinsic (self-contained) geometry of all smooth surfaces, and thereby found a different non-Euclidean geometry. This work of Riemann later became fundamental for Einstein's theory of relativity.

William Blake's "Newton" is a demonstration of his opposition to the 'single-vision' of scientific materialism; here, Isaac Newton is shown as 'divine geometer' (1795)

It remained to be proved mathematically that the non-Euclidean geometry was just as self-consistent as Euclidean geometry, and this was first accomplished by Beltrami in 1868. With this, non-Euclidean geometry was established on an equal mathematical footing with Euclidean geometry.

While it was now known that different geometric theories were mathematically possible, the question remained, "Which one of these theories is correct for our physical space?" The mathematical work revealed that this question must be answered by physical experimentation, not mathematical reasoning, and uncovered the reason why the experimentation must involve immense (interstellar, not earth-bound) distances. With the development of relativity theory in physics, this question became vastly more complicated.

Introduction of Mathematical Rigor

All the work related to the Parallel Postulate revealed that it was quite difficult for a geometer to separate his logical reasoning from his intuitive understanding of physical space, and, moreover, revealed the critical importance of doing so. Careful examination had uncovered some logical inadequacies in Euclid's reasoning, and some unstated geometric principles to which Euclid sometimes appealed. This critique paralleled the crisis occurring in calculus and analysis regarding the meaning of infinite processes such as convergence and continuity. In geometry, there was a clear need for a new set of axioms, which would be complete, and which in no way relied on pictures we draw or on our intuition of space. Such axioms, now known as Hilbert's axioms, were given by David Hilbert in 1894 in his dissertation Grundlagen der Geometrie (Foundations of Geometry). Some other complete sets of axioms had been given a few years earlier, but did not match Hilbert's in economy, elegance, and similarity to Euclid's axioms.

Analysis Situs, or Topology

In the mid-18th century, it became apparent that certain progressions of mathematical reasoning recurred when similar ideas were studied on the number line, in two dimensions, and in three dimensions. Thus the general concept of a metric space was created so that the reasoning could be done in more generality, and then applied to special cases. This method of studying calculus- and analysis-related concepts came to be known as analysis situs, and later as topology. The important topics in this field were properties of more general figures, such as connectedness and boundaries, rather than properties like straightness, and precise equality of length and angle measurements, which had been the focus of Euclidean and non-Euclidean geometry. Topology soon became a separate field of major importance, rather than a sub-field of geometry or analysis.

The 20th Century

Developments in algebraic geometry included the study of curves and surfaces over finite fields as demonstrated by the works of among others André Weil, Alexander Grothendieck, and Jean-Pierre Serre as well as over the real or complex numbers. Finite geometry itself, the study of spaces with only finitely many points, found applications in coding theory and cryptography. With the advent of the computer, new disciplines such as computational geometry or digital geometry deal with geometric algorithms, discrete representations of geometric data, and so forth.

A European and an Arab practicing geometry in the 15th century.

The earliest recorded beginnings of geometry can be traced to ancient Mesopotamia and Egypt in the 2nd millennium BC. Early geometry was a collection of empirically discovered principles concerning lengths, angles, areas, and volumes, which were developed to meet some practical need in surveying, construction, astronomy, and various crafts. The earliest known texts on geometry are the Egyptian Rhind Papyrus (2000–1800 BC) and Moscow Papyrus (c. 1890 BC), the Babylonian clay tablets such as Plimpton 322 (1900 BC). For example, the Moscow Papyrus gives a formula for calculating the volume of a truncated pyramid, or frustum. Later clay tablets (350–50 BC) demonstrate that Babylonian astronomers implemented trapezoid procedures for computing Jupiter's position and motion within time-velocity space. These geometric procedures anticipated the Oxford Calculators, including the mean speed theorem, by 14 centuries. South of Egypt the ancient Nubians established a system of geometry including early versions of sun clocks.

In the 7th century BC, the Greek mathematician Thales of Miletus used geometry to solve problems such as calculating the height of pyramids and the distance of ships from the shore. He is credited with the first use of deductive reasoning applied to geometry, by deriving four corollaries to Thales› Theorem. Pythagoras established the Pythagorean School, which is credited with the first proof of the Pythagorean theorem, though the statement of the theorem has a long history Eudoxus (408–c. 355 BC) developed the method of exhaustion, which allowed the calculation of areas and volumes of curvilinear figures, as well as a theory of ratios that avoided the problem of incommensurable magnitudes, which enabled subsequent geometers to make significant advances. Around 300 BC, geometry was revolutionized by Euclid, whose Elements, widely considered the most successful and influential textbook of all time, introduced mathematical rigor through the axiomatic method and is the earliest example of the format still used in mathematics today, that of definition, axiom, theorem, and proof. Although most of the contents of the Elements were already known, Euclid arranged them into a single, coherent logical framework. The Elements was known to all educated people in the West until the middle of the 20th century and its contents are still taught in geometry classes today.Archimedes (c. 287–212 BC) of Syracuse used the method of exhaustion to calculate the area under the arc of a parabola with the summation of an infinite series, and gave remarkably accurate approximations of Pi. He also studied the spiral bearing his name and obtained formulas for the volumes of surfaces of revolution.

Woman teaching geometry. Illustration at the beginning of a medieval translation of Euclid's Elements, (c. 1310)

Indian mathematicians also made many important contributions in geometry. The Satapatha Brahmana (3rd century BC) contains rules for ritual geometric constructions that are similar to the Sulba Sutras.According to (Hayashi 2005, p. 363), the Śulba Sūtras contain "the earliest extant verbal expression of the Pythagorean Theorem in the world, although it had already been known to the Old Babylonians. They contain lists of Pythagorean triples, which are particular cases of Diophantine equations. In the Bakhshali manuscript, there is a handful of geometric problems (including problems about volumes of irregular solids). The Bakhshali manuscript also employs a decimal place value system with a dot for zero. Aryabhata›s Aryabhatiya (499) includes the computation of areas and volumes. Brahmagupta wrote his astronomical work Brāhma Sphua Siddhānta in 628. Chapter 12, containing 66 Sanskrit verses, was divided into two sections: "basic operations" (including cube roots, fractions, ratio and proportion, and barter) and "practical mathematics" (including mixture, mathematical series, plane figures, stacking bricks, sawing of timber, and piling of grain). In the latter section, he stated his famous theorem on the diagonals of a cyclic quadrilateral. Chapter 12 also included a formula for the area of a cyclic quadrilateral (a generalization of Heron's formula), as well as a complete description of rational triangles (i.e. triangles with rational sides and rational areas).

In the Middle Ages, mathematics in medieval Islam contributed to the development of geometry, especially algebraic geometry. Al-Mahani (b. 853) conceived the idea of reducing geometrical problems such as duplicating the cube to problems in algebra. Thābit ibn Qurra (known as Thebit in Latin) (836–901) dealt with arithmetic operations applied to ratios of geometrical quantities, and contributed to the development of analytic geometry. Omar Khayyám (1048–1131) found geometric solutions to cubic equations. The theorems of Ibn al-Haytham (Alhazen), Omar Khayyam and Nasir al-Din al-Tusi on quadrilaterals, including the Lambert quadrilateral and Saccheri quadrilateral, were early results in hyperbolic geometry, and along with their alternative postulates, such as Playfair›s axiom, these works had a considerable influence on the development of non-Euclidean geometry among later European geometers, including Witelo (c. 1230–c. 1314), Gersonides (1288–1344), Alfonso, John Wallis, and Giovanni Girolamo Saccheri.

In the early 17th century, there were two important developments in geometry. The first was the creation of analytic geometry, or geometry with coordinates and equations, by René Descartes (1596–1650) and Pierre de Fermat (1601–1665). This was a necessary precursor to the development of calculus and a precise quantitative science of physics. The second geometric development of this period was the systematic study of projective geometry by Girard Desargues (1591–1661). Projective geometry is a geometry without measurement or parallel lines, just the study of how points are related to each other.

Two developments in geometry in the 19th century changed the way it had been studied previously. These were the discovery of non-Euclidean geometries by Nikolai Ivanovich Lobachevsky, János Bolyai and Carl Friedrich Gauss and of the formulation of symmetry as the central consideration in the Erlangen Programme of Felix Klein (which generalized the Euclidean and non-Euclidean geometries). Two of the master geometers of the time were Bernhard Riemann (1826–1866), working primarily with tools from mathematical analysis, and introducing the Riemann surface, and Henri Poincaré, the founder of algebraic topology and the geometric theory of dynamical systems. As a consequence of these major changes in the conception of geometry, the concept of space became something rich and varied, and the natural background for theories as different as complex analysis and classical mechanics.

Important concepts in geometry

The following are some of the most important concepts in geometry.

Axioms

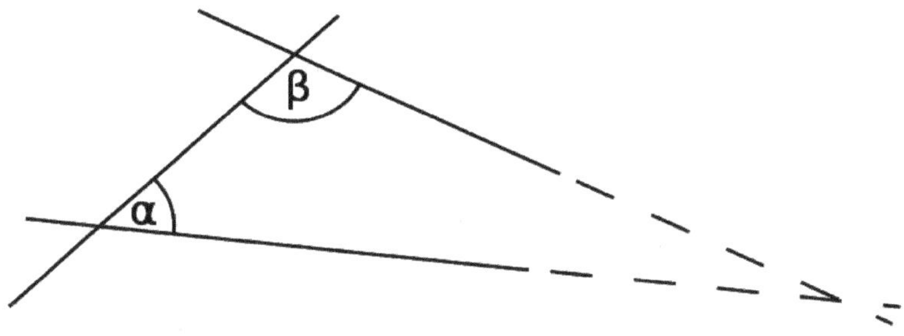

An illustration of Euclid's parallel postulate

Euclid took an abstract approach to geometry in his Elements, one of the most influential books ever written. Euclid introduced certain axioms, or postulates, expressing primary or self-evident properties of points, lines, and planes. He proceeded to rigorously deduce other properties by mathematical reasoning. The characteristic feature of Euclid's approach to geometry was its rigor, and it has come to be known as axiomatic or synthetic geometry. At the start of the 19th century, the discovery of non-Euclidean geometries by Nikolai Ivanovich Lobachevsky (1792–1856), János Bolyai (1802–1860), Carl Friedrich Gauss (1777–1855) and others led to a revival of interest in this discipline, and in the 20th century, David Hilbert (1862–1943) employed axiomatic reasoning in an attempt to provide a modern foundation of geometry.

Points

Points are considered fundamental objects in Euclidean geometry. They have been defined in a variety of ways, including Euclid's definition as 'that which has no part' and through the use of algebra or nested sets. In many areas of geometry, such as analytic geometry, differential geometry, and topology, all objects are considered to be built up from points. However, there has been some study of geometry without reference to points.

Lines

Euclid described a line as "breadthless length" which "lies equally with respect to the points on itself". In modern mathematics, given the multitude of geometries, the concept of a line is closely tied to the way the geometry is described. For instance, in analytic geometry, a line in the plane is often defined as the set of points whose coordinates satisfy a given linear equation, but in a more abstract setting, such as incidence geometry, a line may be an independent object, distinct from the set of points which lie on it. In differential geometry, a geodesic is a generalization of the notion of a line to curved spaces.

Planes

A plane is a flat, two-dimensional surface that extends infinitely far. Planes are used in every area of geometry. For instance, planes can be studied as a topological surface without reference to distances or angles; it can be studied as an affine space, where collinearity and ratios can be studied but not distances; it can be studied as the complex plane using techniques of complex analysis; and so on.

Angles

Euclid defines a plane angle as the inclination to each other, in a plane, of two lines which meet each other, and do not lie straight with respect to each other. In modern terms, an angle is the figure formed by two rays, called the sides of the angle, sharing a common endpoint, called the vertex of the angle.

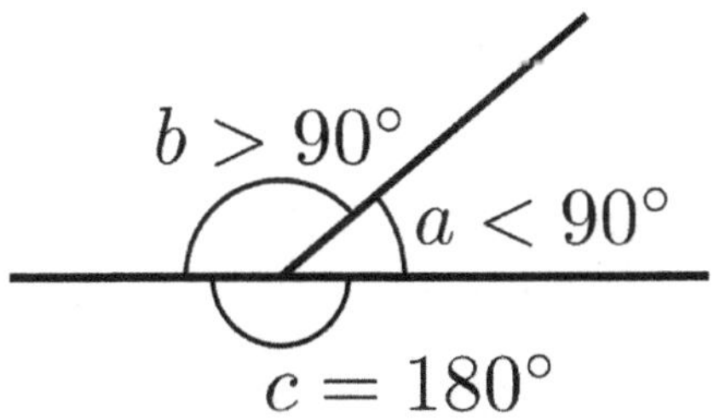

Acute (a), obtuse (b), and straight (c) angles. The acute and obtuse angles are also known as oblique angles.

In Euclidean geometry, angles are used to study polygons and triangles, as well as forming an object of study in their own right. The study of the angles of a triangle or of angles in a unit circle forms the basis of trigonometry.

In differential geometry and calculus, the angles between plane curves or space curves or surfaces can be calculated using the derivative.

Curves

A curve is a 1-dimensional object that may be straight (like a line) or not; curves in 2-dimensional space are called plane curves and those in 3-dimensional space are called space curves.

In topology, a curve is defined by a function from an interval of the real numbers to another space. In differential geometry, the same definition is used, but the defining function is required to be differentiable Algebraic geometry studies algebraic curves, which are defined as algebraic varieties of dimension one.

Surfaces

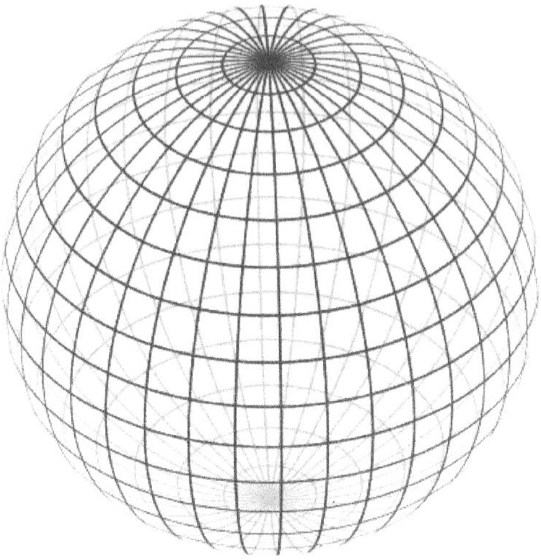

A sphere is a surface that can be defined parametrically (by $x = r \sin \theta \cos \varphi$, $y = r \sin \theta \sin \varphi$, $z = r \cos \theta$) or implicitly (by $x^2 + y^2 + z^2 - r^2 = 0$.)

A surface is a two-dimensional object, such as a sphere or paraboloid. In differential geometry and topology,surfaces are described by two-dimensional ‹patches› (or neighborhoods) that are assembled by diffeomorphisms or homeomorphisms, respectively. In algebraic geometry, surfaces are described by polynomial equations.

Manifolds

A manifold is a generalization of the concepts of curve and surface. In topology, a manifold is a topological space where every point has a neighborhood that is homeomorphic to Euclidean space. In differential geometry, a differentiable manifold is a space where each neighborhood is diffeomorphic to Euclidean space.

Manifolds are used extensively in physics, including in general relativity and string theory

Topologies and Metrics

A topology is a mathematical structure on a set that tells how elements of the set relate spatially to each other. The best-known examples of topologies come from metrics, which are ways of measuring distances between points. For instance, the Euclidean metric measures the distance between points in the Euclidean plane, while the hyperbolic metric measures the distance in the hyperbolic plane. Other important examples of metrics include the Lorentz metric of special relativity and the semi-Riemannian metrics of general relativity.

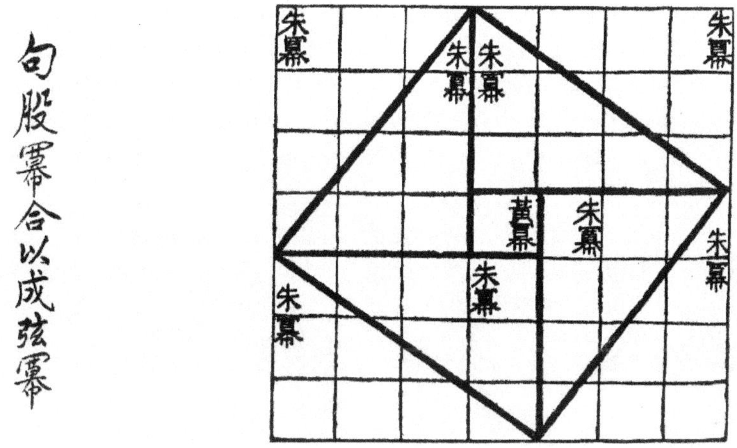

Visual checking of the Pythagorean theorem for the (3, 4, 5) triangle as in the Zhoubi Suanjing 500–200 BC. The Pythagorean theorem is a consequence of the Euclidean metric.

Compass and Straightedge Constructions

Classical geometers paid special attention to constructing geometric objects that had been described in some other way. Classically, the only instruments allowed in geometric constructions are the compass and straightedge. Also, every construction had to be complete in a finite number of steps. However, some problems turned out to be difficult or impossible to solve by these means alone, and ingenious constructions using parabolas and other curves, as well as mechanical devices, were found.

Dimension

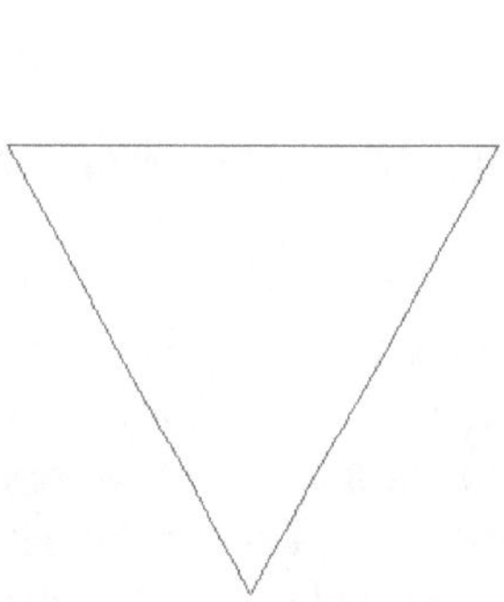

The Koch snowflake, with fractal dimension=log4/log3 and topological dimension=1

Where the traditional geometry allowed dimensions 1 (a line), 2 (a plane) and 3 (our ambient world conceived of as three-dimensional space), mathematicians have used higher dimensions for nearly two centuries. Dimension has gone through stages of being any natural number n, possibly infinite with the introduction of Hilbert space, and any positive real number in fractal geometry. Dimension theory is a technical area, initially within general topology, that discusses definitions; in common with most mathematical ideas, dimension is now defined rather than an intuition. Connected topological manifolds have a well-defined dimension; this is a theorem (invariance of domain) rather than anything a priori.

The issue of dimension still matters to geometry, in the absence of complete answers to classic questions. Dimensions 3 of space and 4 of space-time are special cases in geometric topology. Dimension 10 or 11 is a key number in string theory. Research may bring a satisfactory geometric reason for the significance of 10 and 11 dimensions.

Symmetry

A tiling of the hyperbolic plane

The theme of symmetry in geometry is nearly as old as the science of geometry itself. Symmetric shapes such as the circle, regular polygons and platonic solids held deep significance for many ancient philosophers and were investigated in detail before the time of Euclid. Symmetric patterns occur in nature and were artistically rendered in a multitude of forms, including the graphics of M. C. Escher. Nonetheless, it was not until the second half of 19th century that the unifying role of symmetry in foundations of geometry was recognized. Felix Klein's Erlangen program proclaimed that, in a very precise sense, symmetry, expressed via the notion of a transformation group, determines what geometry is. Symmetry in classical Euclidean geometry is represented by congruences and rigid motions, whereas in projective geometry an analogous role is played by collineations, geometric transformations that take straight lines into straight lines. However it was in the new geometries of Bolyai and Lobachevsky, Riemann, Clifford and Klein, and Sophus Lie that Klein›s idea to ‹define a geometry via its symmetry group› proved most influential. Both discrete and continuous symmetries play prominent roles in geometry, the former in topology and geometric group theory, the latter in Lie theory and Riemannian geometry.

A different type of symmetry is the principle of duality in projective geometry (see Duality (projective geometry)) among other fields. This meta-phenomenon can roughly be described as follows: in any theorem, exchange point with plane, join with meet, lies in with contains, and you will get an equally true theorem. A similar and closely related form of duality exists between a vector space and its dual space.

Non-Euclidean Geometry

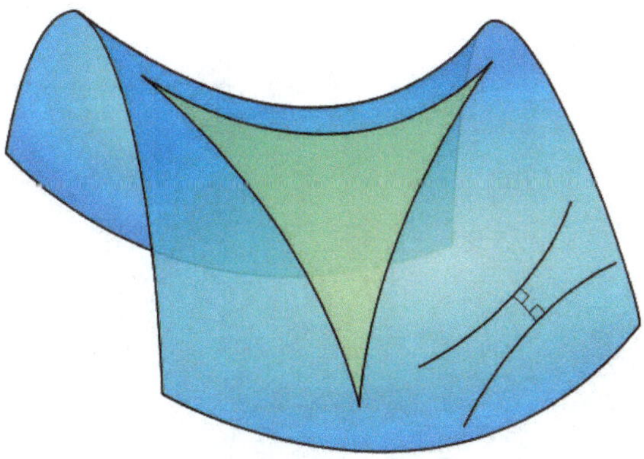

Differential geometry uses tools from calculus to study problems involving curvature.

In the nearly two thousand years since Euclid, while the range of geometrical questions asked and answered inevitably expanded, the basic understanding of space remained essentially the same. Immanuel Kant argued that there is only one, absolute, geometry, which is known to be true a priori by an inner faculty of mind: Euclidean geometry was synthetic a priori. This dominant view was overturned by the revolutionary discovery of non-Euclidean geometry in the works of Bolyai, Lobachevsky, and Gauss (who never published his theory). They demonstrated that ordinary Euclidean space is only one possibility for development of geometry. A broad vision of the subject of geometry was then expressed by Riemann in his 1867 inauguration lecture Über die Hypothesen, welche der Geometrie zu Grunde liegen (On the hypotheses on which geometry is based), published only after his death. Riemann's new idea of space proved crucial in Einstein's general relativity theory, and Riemannian geometry, that considers very general spaces in which the notion of length is defined, is a mainstay of modern geometry.

Contemporary Geometry

Euclidean Geometry

Geometry lessons in the 20th century

Euclidean geometry has become closely connected with computational geometry, computer graphics, convex geometry, incidence geometry, finite geometry, discrete geometry, and some areas of combinatorics. Attention was given to further work on Euclidean geometry and the Euclidean groups by crystallography and the work of H. S. M. Coxeter, and can be seen in theories of Coxeter groups and polytopes. Geometric group theory is an expanding area of the theory of more general discrete groups, drawing on geometric models and algebraic techniques.

Differential Geometry

Differential geometry has been of increasing importance to mathematical physics due to Einstein's general relativity postulation that the universe is curved. Contemporary differential geometry is intrinsic, meaning that the spaces it considers are smooth manifolds whose geometric structure is governed by a Riemannian metric, which determines how distances are measured near each point, and not a priori parts of some ambient flat Euclidean space.

Topology and Geometry

A thickening of the trefoil knot

The field of topology, which saw massive development in the 20th century, is in a technical sense a type of transformation geometry, in which transformations are homeomorphisms. This has often been expressed in the form of the dictum 'topology is rubber-sheet geometry'. Contemporary geometric topology and differential topology, and particular subfields such as Morse theory, would be counted by most mathematicians as part of geometry. Algebraic topology and general topology have gone their own ways.

Algebraic Geometry

Quintic Calabi–Yau threefold

The field of algebraic geometry is the modern incarnation of the Cartesian geometry of co-ordinates. From late 1950s through mid-1970s it had undergone major foundational development, largely due to work of Jean-Pierre Serre and Alexander Grothendieck. This led to the introduction of schemes and greater emphasis on topological methods, including various cohomology theories. One of seven Millennium Prize problems, the Hodge conjecture, is a question in algebraic geometry.

The study of low-dimensional algebraic varieties, algebraic curves, algebraic surfaces and algebraic varieties of dimension 3 (algebraic threefolds), has been far advanced. Gröbner basis theory and real algebraic geometry are among more applied subfields of modern algebraic geometry. Arithmetic geometry is an active field combining algebraic geometry and number theory. Other directions of research involve moduli spaces and complex geometry. Algebro-geometric methods are commonly applied in string and brane theory.

Applications

Geometry has found applications in many fields, some of which are described below.

Art

Mathematics and art are related in a variety of ways. For instance, the theory of perspective showed that there is more to geometry than just the metric properties of figures: perspective is the origin of projective geometry.

Architecture

Mathematics and architecture are related, since, as with other arts, architects use mathematics for several reasons. Apart from the mathematics needed when engineering buildings, architects use geometry: to define the spatial form of a building; from the Pythagoreans of the sixth century BC onwards, to create forms considered harmonious, and thus to lay out buildings and their surroundings according to mathematical, aesthetic and sometimes religious principles; to decorate buildings with mathematical objects such as tessellations; and to meet environmental goals, such as to minimise wind speeds around the bases of tall buildings.

Physics

The field of astronomy, especially as it relates to mapping the positions of stars and planets on the celestial sphere and describing the relationship between movements of celestial bodies, have served as an important source of geometric problems throughout history.

Modern geometry has many ties to physics as is exemplified by the links between pseudo-Riemannian geometry and general relativity. One of the youngest physical theories, string theory, is also very geometric in flavour.

The 4_{21}polytope, orthogonally projected into the E_8 Lie groupCoxeter plane. Lie groups have several applications in physics.

Other Fields of Mathematics

Geometry has also had a large effect on other areas of mathematics. For instance, the introduction of coordinates by René Descartes and the concurrent developments of algebra marked a new stage for geometry, since geometric figures such as plane curves could now be represented analytically in the form of functions and equations. This played a key role in the emergence of infinitesimal calculus in the 17th century. The subject of geometry was further enriched by the study of the intrinsic structure of geometric objects that originated with Euler and Gauss and led to the creation of topology and differential geometry.

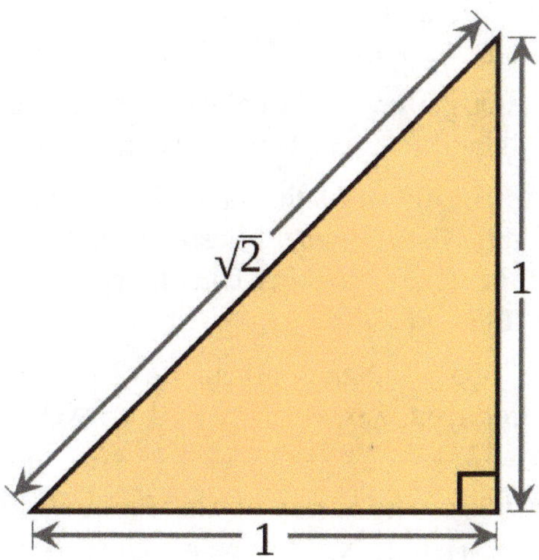

The Pythagoreans discovered that the sides of a triangle could have incommensurable lengths.

An important area of application is number theory. In ancient Greece the Pythagoreans considered the role of numbers in geometry. However, the discovery of incommensurable lengths, which contradicted their philosophical views, made them abandon abstract numbers in favor of concrete geometric quantities, such as length and area of figures. Since the 19th century, geometry has been used for solving problems in number theory, for example through the geometry of numbers or, more recently, scheme theory, which is used in Wiles's proof of Fermat's Last Theorem.

While the visual nature of geometry makes it initially more accessible than other mathematical areas such as algebra or number theory, geometric language is also used in contexts far removed from its traditional, Euclidean provenance (for example, in fractal geometry and algebraic geometry).

Analytic geometry applies methods of algebra to geometric questions, typically by relating geometric curves to algebraic equations. These ideas played a key role in the development of calculus in the 17th century and led to the discovery of many new properties of plane curves. Modern algebraic geometry considers similar questions on a vastly more abstract level.

Leonhard Euler, in studying problems like the Seven Bridges of Königsberg, considered the most fundamental properties of geometric figures based solely on shape, independent of their metric properties. Euler called this new branch of geometry geometria situs (geometry of place), but it is now known as topology. Topology grew out of geometry, but turned into a large independent discipline. It does not differentiate between objects that can be continuously deformed into each other. The objects may nevertheless retain some geometry, as in the case of hyperbolic knots.

REFERENCES

1. Cooke, Roger (2005), The History of Mathematics:, New York: Wiley-Interscience, 632 pages, ISBN 0-471-44459-6
2. Dani, S. G. (July 25, 2003), "On the Pythagorean triples in the Śulvasūtras (PDF), Current Science, 85 (2): 219–224
3. Eder, Michelle (2000), Views of Euclid›s Parallel Postulate in Ancient Greece and in Medieval Islam, Rutgers University, retrieved 2008-01-23
4. Hayashi, Takao (2003), Indian Mathematics, in Grattan-Guinness, Ivor, Companion Encyclopedia of the History and Philosophy of the Mathematical Sciences, 1, Baltimore, MD: The Johns Hopkins University Press, 976 pages, pp. 118–130, ISBN 0-8018-7396-7
5. Hayashi, Takao (2005), Indian Mathematics, in Flood, Gavin, The Blackwell Companion to Hinduism, Oxford: Basil Blackwell, 616 pages, pp. 360–375, ISBN 978-1-4051-3251-0
6. Joseph, G. G. (2000), The Crest of the Peacock: The Non-European Roots of Mathematics, Princeton, NJ: Princeton University Press, 416 pages, ISBN 0-691-00659-8

7. Katz, Victor J. (1998), History of Mathematics: An Introduction, Addison-Wesley, ISBN 0-321-01618-1, OCLC 38199387
8. Needham, Joseph (1986), Science and Civilization in China: Volume 3, Mathematics and the Sciences of the Heavens and the Earth, Taipei: Caves Books Ltd
9. Rozenfeld, Boris A. (1988), A History of Non-Euclidean Geometry: Evolution of the Concept of a Geometric Space, Springer Science+Business Media, ISBN 0-387-96458-4, OCLC 15550634
10. Smith, John D. (1992), The Remarkable Ibn al-Haytham, The Mathematical Gazette, Mathematical Association, 76 (475): 189–198, doi:10.2307/3620392, JSTOR 3620392
11. Staal, Frits (1999), Greek and Vedic Geometry, Journal of Indian Philosophy, 27 (1–2): 105–127, doi:10.1023/A:1004364417713
12. Stillwell, John (2004), Berlin and New York: Mathematics and its History (2 ed.), Springer, 568 pages, ISBN 0-387-95336-1

CHAPTER 2

Concepts and Features of Geometry

Chapter Contents

The dimension of an object is a topological measure of the size of its covering properties. Roughly speaking, it is the number of coordinates needed to specify a point on the object. For example, a rectangle is two-dimensional, while a cube is three-dimensional. The dimension of an object is sometimes also called its "dimensionality."

INTRODUCTION

In physics and mathematics, the **dimension** of a mathematical space (or object) is informally defined as the minimum number of coordinates needed to specify any point within it. Thus a line has a dimension of one because only one coordinate is needed to specify a point on it – for example, the point at 5 on a number line. A surface such as a plane or the surface of a cylinder or sphere has a dimension of two because two coordinates are needed to specify a point on it – for example, both a latitude and longitude are required to locate a point on the surface of a sphere. The inside of a cube, a cylinder or a sphere is three-dimensional because three coordinates are needed to locate a point within these spaces.

In classical mechanics, space and time are different categories and refer to absolute space and time. That conception of the world is a four-dimensional space but not the one that was found necessary to describe electromagnetism. The four dimensions of spacetime consist of events that are not absolutely defined spatially and temporally, but rather are known relative to the motion of an observer. Minkowski space first approximates the universe without gravity; the pseudo-Riemannian manifolds of general relativity describe spacetime with matter and gravity. Ten dimensions are used to describe string theory, and the state-space of quantum mechanics is an infinite-dimensional function space.

The concept of dimension is not restricted to physical objects. **High-dimensional spaces** frequently occur in mathematics and the sciences. They may be parameter spaces or configuration spaces such as in Lagrangian or Hamiltonian mechanics; these are abstract spaces, independent of the physical space we live in.

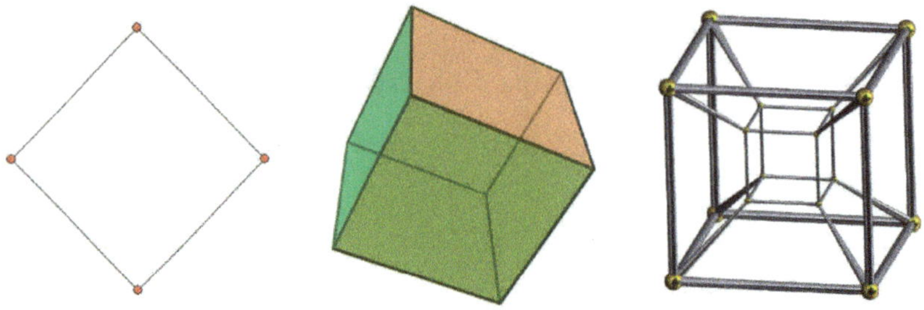

From left to right: the square, the cube and the tesseract. The two-dimensional (2d) square is bounded by one-dimensional (1d) lines; the three-dimensional (3d) cube by two-dimensional areas; and the four-dimensional (4d) tesseract by three-dimensional volumes. For display on a two-dimensional surface such as a screen, the 3d cube and 4d tesseract require projection.

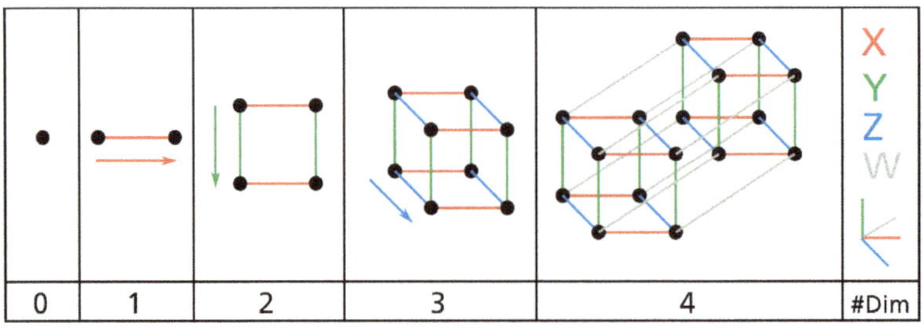

The first four spatial dimensions.

1 Two points can be connected to create a line segment.
2 Two parallel line segments can be connected to form a square.
3 Two parallel squares can be connected to form a cube.
4 Two parallel cubes can be connected to form a tesseract.

IN MATHEMATICS

In mathematics, the dimension of an object is an intrinsic property independent of the space in which the object is embedded. For example, a point on the unit circle in the plane can be specified by two Cartesian coordinates, but a single polar coordinate (the angle) would be sufficient, so the circle is 1-dimensional even though it exists in the 2-dimensional plane. This intrinsic notion of dimension is one of the chief ways the mathematical notion of dimension differs from its common usages.

The dimension of Euclidean n-space $\mathbf{E}^n$ is n. When trying to generalize to other types of spaces, one is faced with the question "what makes $\mathbf{E}^n$ n-dimensional?" One answer is that to cover a fixed ball in $\mathbf{E}^n$ by small balls of radius ε, one needs on the order of ε^{-n} such small balls. This observation leads to the definition of the Minkowski dimension and its more sophisticated variant, the Hausdorff dimension, but there are also other answers to that question. For example, the boundary of a ball in $\mathbf{E}^n$ looks locally like $\mathbf{E}^{n-1}$ and this leads to the notion of the inductive dimension. While these notions agree on $\mathbf{E}^n$, they turn out to be different when one looks at more general spaces.

A tesseract is an example of a four-dimensional object. Whereas outside mathematics the use of the term dimension is as in: A tesseract has four dimensions", mathematicians usually express this as: "The tesseract has dimension 4", or: "The dimension of the tesseract is 4".

Although the notion of higher dimensions goes back to René Descartes, substantial development of a higher-dimensional geometry only began in the 19th century, via the work of Arthur Cayley, William Rowan Hamilton, Ludwig Schläfli and Bernhard Riemann. Riemann's 1854 Habilitationsschrift, Schläfli's 1852 Theorie der vielfachen Kontinuität, Hamilton's 1843 discovery of the quaternions and the construction of the Cayley algebra marked the beginning of higher-dimensional geometry.

The rest of this section examines some of the more important mathematical definitions of the dimensions.

Complex Dimension

A complex number (x + iy) has a real part x and an imaginary part y, where x and y are real numbers. A single complex coordinate system may be applied to an object having two real dimensions. For example, an ordinary two-dimensional spherical surface, when given a complex metric, becomes a Riemann sphere of one complex dimension. Complex dimensions appear in the study of complex manifolds and algebraic varieties.

Dimension of a Vector Space

The dimension of a vector space is the number of vectors in any basis for the space, i.e. the number of coordinates necessary to specify any vector. This notion of dimension (the cardinality of a basis) is often referred to as the Hamel dimension or algebraic dimension to distinguish it from other notions of dimension.

Manifolds

A connected topological manifold is locally homeomorphic to Euclidean n-space, and the number n is called the manifold's dimension. One can show that this yields a uniquely defined dimension for every connected topological manifold.

For connected differentiable manifolds, the dimension is also the dimension of the tangent vector space at any point.

In geometric topology, the theory of manifolds is characterized by the way dimensions 1 and 2 are relatively elementary, the **high-dimensional** cases $n > 4$ are simplified by having extra space in which to "work"; and the cases $n = 3$ and 4 are in some senses the most difficult. This state of affairs was highly marked in the various cases of the Poincaré conjecture, where four different proof methods are applied.

Varieties

The dimension of an algebraic variety may be defined in various equivalent ways. The most intuitive way is probably the dimension of the tangent space at any regular point. Another intuitive way is to define the dimension as the number of hyperplanes that are needed in order to have an intersection with the variety that is reduced to a finite number of points (dimension zero). This definition is based on the fact that the intersection of a variety with a hyperplane reduces the dimension by one unless if the hyperplane contains the variety.

An algebraic set being a finite union of algebraic varieties, its dimension is the maximum of the dimensions of its components. It is equal to the maximal length of the chains $V_0 \subsetneq V_1 \subsetneq \cdots \subsetneq V_d$ of sub-varieties of the given algebraic set (the length of such a chain is the number of "$\subsetneq$"). Each variety can be considered as an algebraic stack, and its dimension as variety agrees with its dimension as stack. There are however many stacks which do not correspond to varieties, and some of these have negative dimension. Specifically, if V is a variety of dimension m and G is an algebraic group of dimension n acting on V, then the quotient stack [V/G] has dimension m−n.

Krull Dimension

The Krull dimension of a commutative ring is the maximal length of chains of prime ideals in it, a chain of length n being a sequence $\mathcal{P}_0 \subsetneq \mathcal{P}_1 \subsetneq \cdots \subsetneq \mathcal{P}_n$ of prime ideals related by inclusion. It is strongly related to the dimension of an algebraic variety, because of the natural correspondence between sub-varieties and prime ideals of the ring of the polynomials on the variety.

For an algebra over a field, the dimension as vector space is finite if and only if its Krull dimension is 0.

Lebesgue Covering Dimension

For any normal topological space X, the Lebesgue covering dimension of X is defined to be n if n is the smallest integer for which the following holds: any open cover has an open refinement (a second open cover where each element is a subset of an element in the first cover) such that no point is included in more than n + 1 elements. In this case dim X = n. For X a manifold, this coincides with the dimension mentioned above. If no such integer n exists, then the dimension of X is said to be infinite, and one writes dim X = ∞. Moreover, X has dimension −1, i.e. dim X = −1 if and only if X is empty. This definition of covering dimension can be extended from the class of normal spaces to all Tychonoff spaces merely by replacing the term open in the definition by the term **functionally open**".

Inductive Dimension

An inductive definition of dimension can be created as follows. Consider a discrete set of points (such as a finite collection of points) to be 0-dimensional. By dragging a 0-dimensional object in some direction, one obtains a 1-dimensional object. By dragging a 1-dimensional object in a new direction, one obtains a 2-dimensional object. In general one obtains an (n + 1)-dimensional object by dragging an n-dimensional object in a new direction.

The inductive dimension of a topological space may refer to the small inductive dimension or the large inductive dimension, and is based on the analogy that (n + 1)-dimensional balls have n-dimensional boundaries, permitting an inductive definition based on the dimension of the boundaries of open sets.

Hausdorff Dimension

For structurally complicated sets, especially fractals, the Hausdorff dimension is useful. The Hausdorff dimension is defined for all metric spaces and, unlike the dimensions considered above, can also attain non-integer real values. The box dimension or Minkowski dimension is a variant of the same idea. In general, there exist more definitions of fractal dimensions that work for highly irregular sets and attain non-integer positive real values. Fractals have been found useful to describe many natural objects and phenomena.

Hilbert Spaces

Every Hilbert space admits an orthonormal basis, and any two such bases for a particular space have the same cardinality. This cardinality is called the dimension of the Hilbert space. This dimension is finite if and only if the space's Hamel dimension is finite, and in this case the above dimensions coincide.

In Physics

Spatial Dimensions

Classical physics theories describe three physical dimensions: from a particular point in space, the basic directions in which we can move are up/down, left/right, and forward/ backward. Movement in any other direction can be expressed in terms of just these three. Moving down is the same as moving up a negative distance. Moving diagonally upward and forward is just as the name of the direction implies; i.e., moving in a linear combination of up and forward. In its simplest form: a line describes one dimension, a plane describes two dimensions, and a cube describes three dimensions. (See Space and Cartesian coordinate system.)

<table>
<tr><th>Number of dimensions</th><th colspan="3">Example co-ordinate systems</th></tr>
<tr><td>1</td><td>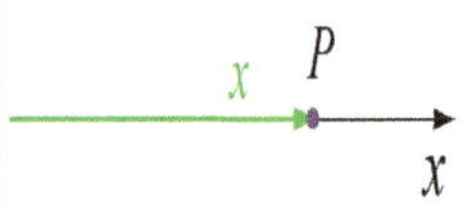

Number line</td><td colspan="2">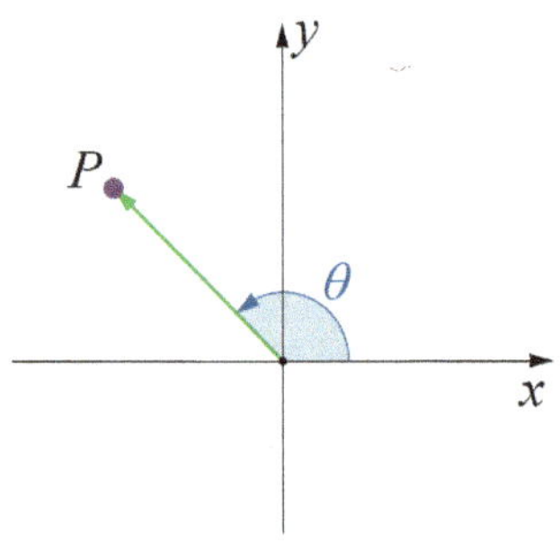

Angle</td></tr>
<tr><td>2</td><td>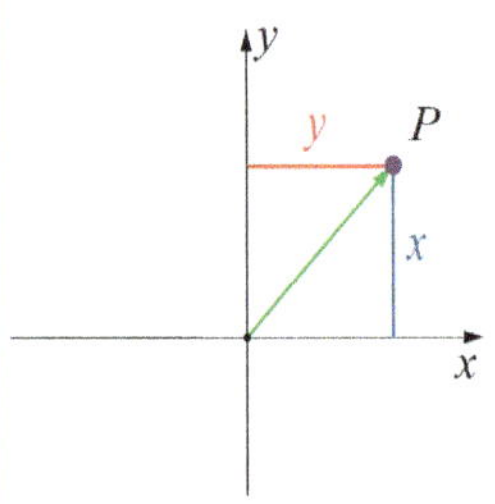

Cartesian (two-dimensional)</td><td>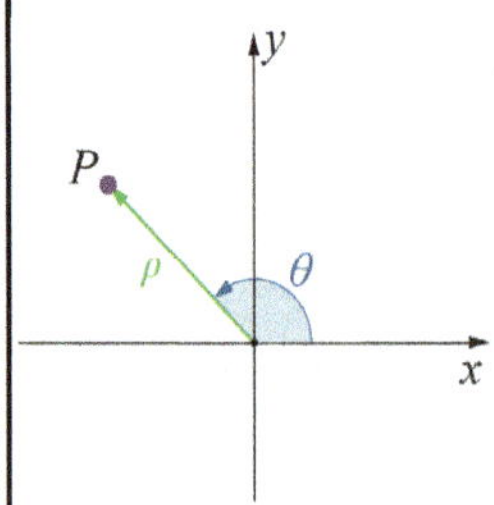

Polar</td><td>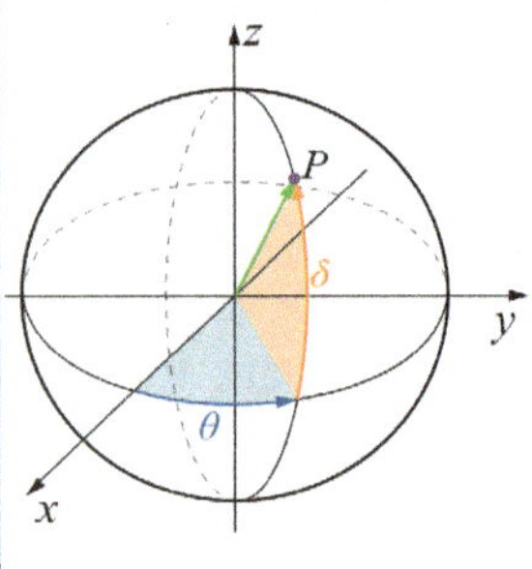

Latitude and longitude</td></tr>
</table>

3	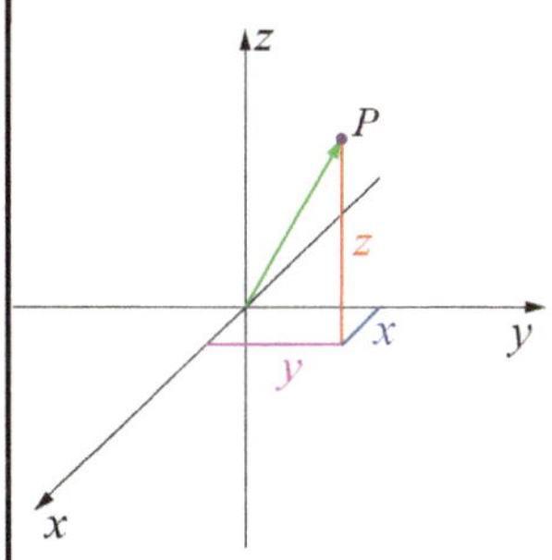 **Cartesian** (three-dimensional)	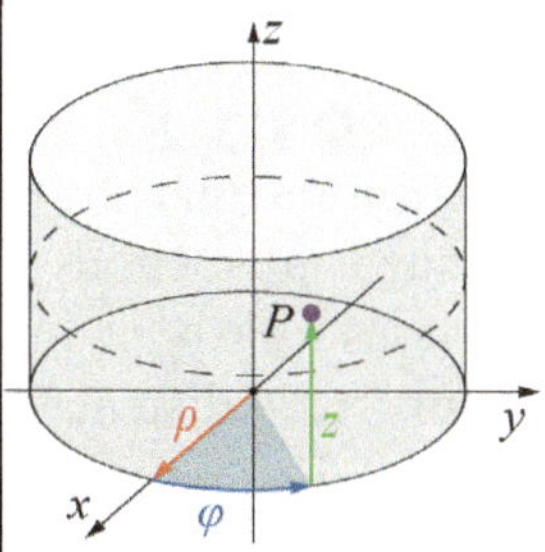 **Cylindrical**	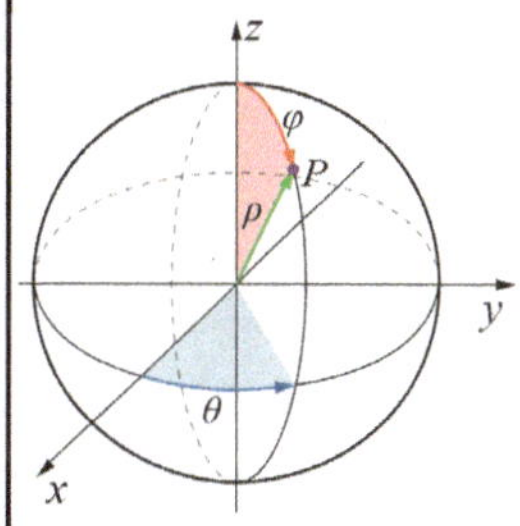 **Spherical**

Time

A temporal dimension is a dimension of time. Time is often referred to as the "fourth dimension" for this reason, but that is not to imply that it is a spatial dimension. A temporal dimension is one way to measure physical change. It is perceived differently from the three spatial dimensions in that there is only one of it, and that we cannot move freely in time but subjectively move in one direction.

The equations used in physics to model reality do not treat time in the same way that humans commonly perceive it. The equations of classical mechanics are symmetric with respect to time, and equations of quantum mechanics are typically symmetric if both time and other quantities (such as charge and parity) are reversed. In these models, the perception of time flowing in one direction is an artifact of the laws of thermodynamics (we perceive time as flowing in the direction of increasing entropy).

The best-known treatment of time as a dimension is Poincaré and Einstein's special relativity (and extended to general relativity), which treats perceived space and time as components of a four-dimensional manifold, known as spacetime, and in the special, flat case as Minkowski space.

Additional Dimensions

"Extradimensional" redirects here. For an album, see Extra Dimensional.

In physics, three dimensions of space and one of time is the accepted norm. However, there are theories that attempt to unify the four fundamental forces by introducing extra dimensions. Most notably, superstring theory requires 10 spacetime dimensions, and originates from a more fundamental 11-dimensional theory tentatively called M-theory which subsumes five previously distinct superstring theories. To date, no experimental or observational evidence is available to confirm the existence of these extra dimensions. If extra dimensions exist, they must be hidden from us by some physical mechanism. One well-studied possibility is that the extra dimensions may be "curled up" at such tiny scales as to be effectively invisible to current experiments. Limits on the size and other properties of extra dimensions are set by particle experiments such as those at the Large Hadron Collider.

At the level of quantum field theory, Kaluza–Klein theory unifies gravity with gauge interactions, based on the realization that gravity propagating in small, compact extra dimensions is equivalent to gauge interactions at long distances. In particular when the geometry of the extra dimensions is trivial, it reproduces electromagnetism. However at sufficiently high energies or short distances, this setup still suffers from the same pathologies that famously obstruct direct attempts to describe quantum gravity. Therefore, these models still require a UV completion, of the kind that string theory is intended to provide. In particular, superstring theory requires six compact dimensions forming a Calabi–Yau manifold. Thus Kaluza-Klein theory may be considered either as an incomplete description on its own, or as a subset of string theory model building.

In addition to small and curled up extra dimensions, there may be extra dimensions that instead aren't apparent because the matter associated with our visible universe is localized on a (3 + 1)-dimensional subspace. Thus the extra dimensions need not be small and compact but may be large extra dimensions. D-branes are dynamical extended objects of various dimensionalities predicted by string theory that could play this role. They have the property that open string excitations, which are associated with gauge interactions, are confined to the brane by their endpoints, whereas the closed strings that mediate the gravitational interaction are free to propagate into the whole spacetime, or "the bulk". This could be related to why gravity is exponentially weaker than the other forces, as it effectively dilutes itself as it propagates into a higher-dimensional volume.

Some aspects of brane physics have been applied to cosmology. For example, brane gas cosmology attempts to explain why there are three dimensions of space using topological and thermodynamic considerations. According to this idea it would be because three is the largest number of spatial dimensions where strings can generically intersect. If initially there are lots of windings of strings around compact dimensions, space could only expand to macroscopic sizes once these windings are eliminated, which requires oppositely wound strings to find each other and annihilate. But strings can only find each other to annihilate at a meaningful rate in three dimensions, so it follows that only three dimensions of space are allowed to grow large given this kind of initial configuration.

Extra dimensions are said to be universal if all fields are equally free to propagate within them.

Networks and Dimension

Some complex networks are characterized by fractal dimensions. The concept of dimension can be generalized to include networks embedded in space. The dimension characterize their spatial constraints.

In Literature

Science fiction texts often mention the concept of "dimension" when referring to parallel or alternate universes or other imagined planes of existence. This usage is derived from the idea that to travel to parallel/alternate universes/planes of existence one must travel in a direction/dimension besides the standard ones. In effect, the other universes/planes are just a small distance away from our own, but the distance is in a fourth (or higher) spatial (or non-spatial) dimension, not the standard ones.

One of the most heralded science fiction stories regarding true geometric dimensionality, and often recommended as a starting point for those just starting to investigate such matters, is the 1884 novella Flatland by Edwin A. Abbott. Isaac Asimov, in his foreword to the Signet Classics 1984 edition, described Flatland as "The best introduction one can find into the manner of perceiving dimensions."

The idea of other dimensions was incorporated into many early science fiction stories, appearing prominently, for example, in Miles J. Breuer's The Appendix and the Spectacles (1928) and Murray Leinster's The Fifth-Dimension Catapult (1931); and appeared irregularly in science fiction by the 1940s. Classic stories involving other dimensions include Robert A. Heinlein's —And He Built a Crooked House (1941), in which a California architect designs a house based on a three-dimensional projection of a tesseract; and Alan E. Nourse's Tiger by the Tail and The Universe Between (both 1951). Another reference is Madeleine L'Engle's novel A Wrinkle In Time (1962), which uses the fifth dimension as a way for "tesseracting the universe or folding space in order to move across it quickly. The fourth and fifth dimensions were also a key component of the book The Boy Who Reversed Himself by William Sleator.

In Philosophy

Immanuel Kant, in 1783, wrote: "That everywhere space (which is not itself the boundary of another space) has three dimensions and that space in general cannot have more dimensions is based on the proposition that not more than three lines can intersect at right angles in one point. This proposition cannot at all be shown from concepts, but rests immediately on intuition and indeed on pure intuition a priori because it is apodictically (demonstrably) certain."

Space has Four Dimensions is a short story published in 1846 by German philosopher and experimental psychologist Gustav Fechner under the pseudonym Dr. Mises. The protagonist in the tale is a shadow who is aware of and able to communicate with other shadows, but who is trapped on a two-dimensional surface. According to Fechner, this shadow-man would conceive of the third dimension as being one of time. The story bears a strong similarity to the Allegory of the Cave presented in Plato's The Republic (c. 380 BC).

Simon Newcomb wrote an article for the Bulletin of the American Mathematical Society in 1898 entitled "The Philosophy of Hyperspace". Linda Dalrymple Henderson coined the term "hyperspace philosophy", used to describe writing that uses higher dimensions to explore metaphysical themes, in her 1983 thesis about the fourth dimension in early-twentieth-century art. Examples of "hyperspace philosophers" include Charles Howard Hinton, the first writer, in 1888, to use the word "tesseract"; and the Russian esotericist P. D. Ouspensky.

COMPASS-AND-STRAIGHTEDGE CONSTRUCTION

Compass-and-straightedge construction, also known as ruler-and-compass construction or classical construction, is the construction of lengths, angles, and other geometric figures using only an idealizedruler and compass.

The idealized ruler, known as a straightedge, is assumed to be infinite in length, and has no markings on it and only one edge. The compass is assumed to collapse when lifted from the page, so may not be directly used to transfer distances. (This is an unimportant restriction since, using a multi-step procedure, a distance can be transferred even with collapsing compass; see compass equivalence theorem.) More formally, the only permissible constructions are those granted by Euclid's first three postulates.

It turns out to be the case that every point constructible using straightedge and compass may also be constructed using compass alone.

The ancient Greek mathematicians first conceived compass-and-straightedge constructions, and a number of ancient problems in plane geometry impose this restriction. The ancient Greeks developed many constructions, but in some cases were unable to do so. Gauss showed that some polygons are constructible but that most are not. Some of the most famous straightedge-and-compass problems were proven impossible by Pierre Wantzel in 1837, using the mathematical theory of fields.

In spite of existing proofs of impossibility, some persist in trying to solve these problems. Many of these problems are easily solvable provided that other geometric transformations are allowed: for example, doubling the cube is possible using geometric constructions, but not possible using straightedge and compass alone.

In terms of algebra, a length is constructible if and only if it represents a constructible number, and an angle is constructible if and only if its cosine is a constructible number. A number is constructible if and only if it can be written using the four basic arithmetic operations and the extraction of square roots but of no higher-order roots.

Compass and Straightedge Tools

The "compass" and "straightedge" of compass and straightedge constructions are idealizations of rulers and compasses in the real world:

- The **compass** can be opened arbitrarily wide, but (unlike some real compasses) it has no markings on it. Circles can only be drawn starting from two given points: the centre and a point on the circle. The compass may or may not collapse when it's not drawing a circle.
- The **straightedge** is infinitely long, but it has no markings on it and has only one straight edge, unlike ordinary rulers. It can only be used to draw a line segment between two points or to extend an existing segment.

The modern compass generally does not collapse and several modern constructions use this feature. It would appear that the modern compass is a more powerful instrument than the ancient collapsing compass. However, by Proposition 2 of Book 1 of Euclid's Elements, no power is lost by using a collapsing compass. Although the proposition is correct, its proofs have a long and checkered history.

Each construction must be exact. "Eyeballing" it (essentially looking at the construction and guessing at its accuracy, or using some form of measurement, such as the units of measure on a ruler) and getting close does not count as a solution.

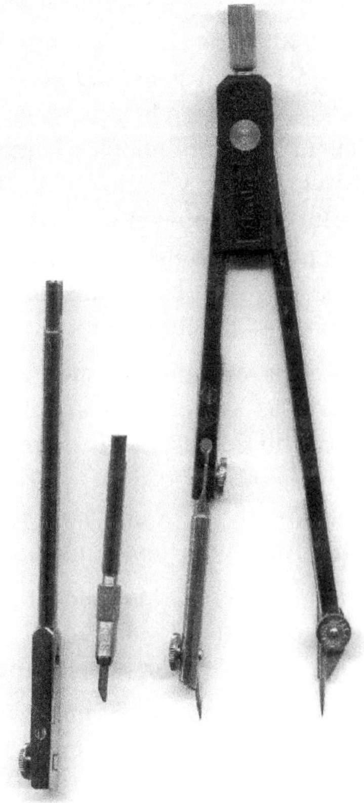

A compass

Each construction must terminate. That is, it must have a finite number of steps, and not be the limit of ever closer approximations.

Stated this way, compass and straightedge constructions appear to be a parlour game, rather than a serious practical problem; but the purpose of the restriction is to ensure that constructions can be proven to be exactly correct, and is thus important to both drafting (design by both CAD software and traditional drafting with pencil, paper, straight-edge and compass) and the science of weights and measures, in which exact synthesis from reference bodies or materials is extremely important. One of the chief purposes of Greek mathematics was to find exact constructions for various lengths; for example, the side of a pentagon inscribed in a given circle. The Greeks could not find constructions for these three problems, among others:

- **Squaring the circle**: Drawing a square the same area as a given circle.
- **Doubling the cube**: Drawing a cube with twice the volume of a given cube.

- **Trisecting the angle**: Dividing a given angle into three smaller angles all of the same size.

For 2000 years people tried to find constructions within the limits set above, and failed. All three have now been proven under mathematical rules to be generally impossible (angles with certain values can be trisected, but not all possible angles).

The ancient Greek mathematicians first attempted compass-and-straightedge constructions, and they discovered how to construct sums, differences, products, ratios, and square roots of given lengths. They could also construct half of a given angle, a square whose area is twice that of another square, a square having the same area as a given polygon, and a regular polygon with 3, 4, or 5 sides (or one with twice the number of sides of a given polygon). But they could not construct one third of a given angle except in particular cases, or a square with the same area as a given circle, or a regular polygon with other numbers of sides. Nor could they construct the side of a cube whose volume would be twice the volume of a cube with a given side.

Hippocrates and Menaechmus showed that the area of the cube could be doubled by finding the intersections of hyperbolas and parabolas, but these cannot be constructed by compass and straightedge. In the fifth century BCE, Hippias used a curve that he called a quadratrix to both trisect the general angle and square the circle, and Nicomedes in the second century BCE showed how to use a conchoid to trisect an arbitrary angle; but these methods also cannot be followed with just compass and straightedge.

No progress on the unsolved problems was made for two millennia, until in 1796 Gauss showed that a regular polygon with 17 sides could be constructed; five years later he showed the sufficient criterion for a regular polygon of n sides to be constructible.

In 1837 Pierre Wantzel published a proof of the impossibility of trisecting an arbitrary angle or of doubling the volume of a cube, based on the impossibility of constructing cube roots of lengths. He also showed that Gauss's sufficient constructibility condition for regular polygons is also necessary.

Then in 1882 Lindemann showed that is a transcendental number, and thus that it is impossible by straightedge and compass to construct a square with the same area as a given circle.

The Basic Constructions

All compass and straightedge constructions consist of repeated application of five basic constructions using the points, lines and circles that have already been constructed. These are:

- Creating the line through two existing points
- Creating the circle through one point with centre another point
- Creating the point which is the intersection of two existing, non-parallel lines
- Creating the one or two points in the intersection of a line and a circle (if they intersect)
- Creating the one or two points in the intersection of two circles (if they intersect).

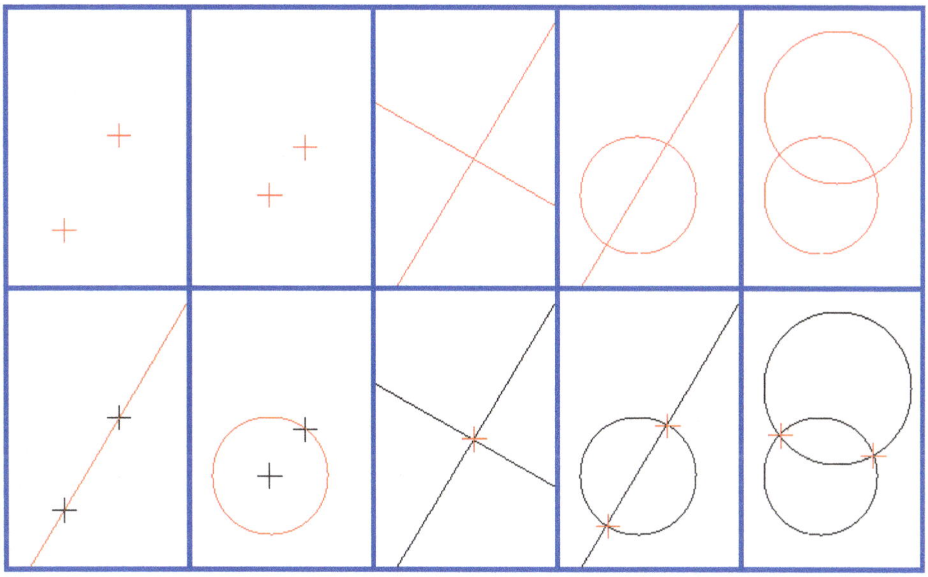

The basic constructions

For example, starting with just two distinct points, we can create a line or either of two circles (in turn, using each point as centre and passing through the other point). If we draw both circles, two new points are created at their intersections. Drawing lines between the two original points and one of these new points completes the construction of an equilateral triangle.

Therefore, in any geometric problem we have an initial set of symbols (points and lines), an algorithm, and some results. From this perspective, geometry is equivalent to an axiomatic algebra, replacing its elements by symbols. Probably Gauss first realized this, and used it to prove the impossibility of some constructions; only much later did Hilbert find a complete set of axioms for geometry.

Much used Compass-and-Straightedge Constructions

The most-used compass-and-straightedge constructions include:

- Constructing the perpendicular bisector from a segment
- Finding the midpoint of a segment.
- Drawing a perpendicular line from a point to a line.
- Bisecting an angle
- Mirroring a point in a line
- Constructing a line through a point tangent to a circle
- Constructing a circle through 3 noncollinear points

Constructible Points and Lengths

Formal Proof

There are many different ways to prove something is impossible. A more rigorous proof would be to demarcate the limit of the possible, and show that to solve these problems one must transgress that limit. Much of what can be constructed is covered in intercept theory.

We could associate an algebra to our geometry using a Cartesian coordinate system made of two lines, and represent points of our plane by vectors. Finally we can write these vectors as complex numbers.

Using the equations for lines and circles, one can show that the points at which they intersect lie in a quadratic extension of the smallest field F containing two points on the line, the center of the circle, and the radius of the circle. That is, they are of the form $x + y\sqrt{k}$, where x, y, and k are in F.

Since the field of constructible points is closed under square roots, it contains all points that can be obtained by a finite sequence of quadratic extensions of the field of complex numbers with rational coefficients. By the above paragraph, one can show that any constructible point can be obtained by such a sequence of extensions. As a corollary of this, one finds that the degree of the minimal polynomial for a constructible point (and therefore of any constructible length) is a power of 2. In particular, any constructible point (or length) is an algebraic number, though not every algebraic number is constructible; for example, $\sqrt[3]{2}$ is algebraic but not constructible.

Constructible Angles

There is a bijection between the angles that are constructible and the points that are constructible on any constructible circle. The angles that are constructible form an abelian group under addition modulo 2π (which corresponds to multiplication of the points on the unit circle viewed as complex numbers). The angles that are constructible are exactly those whose tangent (or equivalently, sine or cosine) is constructible as a number. For example, the regular heptadecagon (the seventeen-sided regular polygon) is constructible because

$$\cos\left(\frac{2\pi}{17}\right) = -\frac{1}{16} + \frac{1}{16}\sqrt{17} + \frac{1}{16}\sqrt{34 - 2\sqrt{17}} + \frac{1}{8}\sqrt{17 + 3\sqrt{17} - \sqrt{34 - 2\sqrt{17}} - 2\sqrt{34 + 2\sqrt{17}}}$$

as discovered by Gauss.

The group of constructible angles is closed under the operation that halves angles (which corresponds to taking square roots in the complex numbers). The only angles of finite order that may be constructed starting with two points are those whose order is either a power of two, or a product of a power of two and a set of distinct Fermat primes. In addition there is a dense set of constructible angles of infinite order.

Compass and Straightedge Constructions as Complex Arithmetic

Given a set of points in the Euclidean plane, selecting any one of them to be called **0** and another to be called **1**, together with an arbitrary choice of orientation allows us to consider the points as a set of complex numbers.

Given any such interpretation of a set of points as complex numbers, the points constructible using valid compass and straightedge constructions alone are precisely the elements of the smallest field containing the original set of points and closed under the complex conjugate and square root operations (to avoid ambiguity, we can specify the square root with complex argument less than π). The elements of this field are precisely those that may be expressed as a formula in the original points using only the operations of addition, subtraction, multiplication, division, complex conjugate, and square root, which is easily seen to be a countable dense subset of the plane. Each of these six operations corresponding to a simple compass and straightedge construction. From such a formula it is straightforward to produce a construction of the corresponding point by combining the constructions for each of the arithmetic operations. More efficient constructions of a particular set of points correspond to shortcuts in such calculations.

Equivalently (and with no need to arbitrarily choose two points) we can say that, given an arbitrary choice of orientation, a set of points determines a set of complex ratios given by the ratios of the differences between any two pairs of points. The set of ratios constructible using compass and straightedge from such a set of ratios is precisely the smallest field containing the original ratios and closed under taking complex conjugates and square roots.

For example, the real part, imaginary part and modulus of a point or ratio z (taking one of the two viewpoints above) are constructible as these may be expressed as

$$\mathrm{Re}(z) = \frac{z+\bar{z}}{2}$$

$$\mathrm{Im}(z) = \frac{z-\bar{z}}{2i}$$

$$|z| = \sqrt{z\bar{z}}.$$

Doubling the cube and trisection of an angle (except for special angles such as any φ such that $\varphi/2\pi$ is a rational number with denominator not divisible by 3) require ratios which are the solution to cubic equations, while squaring the circle requires a transcendental ratio. None of these are in the fields described, hence no compass and straightedge construction for these exists.

Impossible Constructions

The ancient Greeks thought that the construction problems they could not solve were simply obstinate, not unsolvable. With modern methods, however, these compass-and-straightedge constructions have been shown to be logically impossible to perform. (The problems themselves, however, are solvable, and the Greeks knew how to solve them, without the constraint of working only with straightedge and compass.)

Squaring the Circle

The most famous of these problems, **squaring the circle**, otherwise known as the **quadrature of the circle**, involves constructing a square with the same area as a given circle using only straightedge and compass.

Squaring the circle has been proven impossible, as it involves generating a transcendental number, that is, $\sqrt{\pi}$. Only certain algebraic numbers can be constructed with ruler and compass alone, namely those constructed from the integers with a finite sequence of operations of addition, subtraction, multiplication, division, and taking square roots. The phrase "squaring the circle" is often used to mean "doing the impossible" for this reason.

Without the constraint of requiring solution by ruler and compass alone, the problem is easily solvable by a wide variety of geometric and algebraic means, and was solved many times in antiquity.

Doubling the Cube

Doubling the cube is the construction, using only a straight-edge and compass, of the edge of a cube that has twice the volume of a cube with a given edge. This is impossible because the cube root of 2, though algebraic, cannot be computed from integers by addition, subtraction, multiplication, division, and taking square roots. This follows because its minimal polynomial over the rationals has degree 3. This construction is possible using a straightedge with two marks on it and a compass.

Angle Trisection

Angle trisection is the construction, using only a straightedge and a compass, of an angle that is one-third of a given arbitrary angle. This is impossible in the general case. For example, though the angle of $\pi/3$ radians (60°) cannot be trisected, the angle $2\pi/5$ radians (72° = 360°/5) can be trisected. The general trisection problem is also easily solved when a straightedge with two marks on it is allowed (a neusis construction).

Constructing Regular Polygons

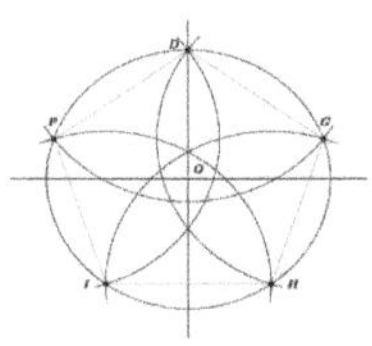

Construction of a regular pentagon

Some regular polygons (e.g. a pentagon) are easy to construct with straightedge and compass; others are not. This led to the question: Is it possible to construct all regular polygons with straightedge and compass?

Carl Friedrich Gauss in 1796 showed that a regular 17-sided polygon can be constructed, and five years later showed that a regular n-sided polygon can be constructed with straightedge and compass if the odd prime factors of n are distinct Fermat primes. Gauss conjectured that this condition was also necessary, but he offered no proof of this fact, which was provided by Pierre Wantzel in 1837.

The first few constructible regular polygons have the following numbers of sides:

3, 4, 5, 6, 8, 10, 12, 15, 16, 17, 20, 24, 30, 32, 34, 40, 48, 51, 60, 64, 68, 80, 85, 96, 102, 120, 128, 136, 160, 170, 192, 204, 240, 255, 256, 257, 272... (sequence A003401 in the OEIS)

There are known to be an infinitude of constructible regular polygons with an even number of sides (because if a regular n-gon is constructible, then so is a regular 2n-gon and hence a regular 4n-gon, 8n-gon, etc.). However, there are only 31 known constructible regular n-gons with an odd number of sides.

Constructing a Triangle from Three Given Characteristic Points or Lengths

Sixteen key points of a triangle are its vertices, the midpoints of its sides, the feet of its altitudes, the feet of its internal angle bisectors, and its circumcenter, centroid, orthocenter, and incenter. These can be taken three at a time to yield 139 distinct nontrivial problems of constructing a triangle from three points. Of these problems, three involve a point that can be uniquely constructed from the other two points; 23 can be non-uniquely constructed (in fact for infinitely many solutions) but only if the locations of the points obey certain constraints; in 74 the problem is constructible in the general case; and in 39 the required triangle exists but is not constructible.

Twelve key lengths of a triangle are the three side lengths, the three altitudes, the three medians, and the three angle bisectors. Together with the three angles, these give 95 distinct combinations, 63 of which give rise to a constructible triangle, 30 of which do not, and two of which are underdefined.

Distance to an Ellipse

The line segment from any point in the plane to the nearest point on a circle can be constructed, but the segment from any point in the plane to the nearest point on an ellipse of positive eccentricity cannot in general be constructed.

Constructing with only Ruler or only Compass

It is possible (according to the Mohr–Mascheroni theorem) to construct anything with just a compass if it can be constructed with a ruler and compass, provided that the given data and the data to be found consist of discrete points (not lines or circles). It should be noted that the truth of this theorem depends on the truth of Archimedes' axiom, which is not first-order in nature. It is impossible to take a square root with just a ruler, so some things that cannot be constructed with a ruler can be constructed with a compass; but (by the Poncelet–Steiner theorem) given a single circle and its center, they can be constructed.

Extended Constructions

The ancient Greeks classified constructions into three major categories, depending on the complexity of the tools required for their solution. If a construction used only a straightedge and compass, it was called planar; if it also required one or more conic sections (other than the circle), then it was called solid; the third category included all constructions that did not fall into either of the other two categories. This categorization meshes nicely with our modern algebraic point of view. A complex number that can be expressed using only the field operations and square roots (as described above) has a planar construction. A complex number that includes also the extraction of cube roots has a solid construction.

In the language of fields, a complex number that is planar has degree a power of two, and lies in a field extension that can be broken down into a tower of fields where each extension has degree two. A complex number that has a solid construction has degree with prime factors of only two and three, and lies in a field extension that is at the top of a tower of fields where each extension has degree 2 or 3.

Solid Constructions

A point has a solid construction if it can be constructed using a straightedge, compass, and a (possibly hypothetical) conic drawing tool that can draw any conic with already constructed focus, directrix, and eccentricity. The same set of points can often be constructed using a smaller set of tools. For example, using a compass, straightedge, and a piece of paper on which we have the parabola $y=x^2$ together with the points (0,0) and (1,0), one can construct any complex number that has a solid construction. Likewise, a tool that can draw any ellipse with already constructed foci and major axis (think two pins and a piece of string) is just as powerful.

The ancient Greeks knew that doubling the cube and trisecting an arbitrary angle both had solid constructions. Archimedes gave a solid construction of the regular 7-gon. The quadrature of the circle does not have a solid construction.

A regular n-gon has a solid construction if and only if $n=2^j3^km$ where m is a product of distinct Pierpont primes (primes of the form 2^r3^s+1). The set of such n is the sequence

7, 9, 13, 14, 18, 19, 21, 26, 27, 28, 35, 36, 37, 38, 39, 42, 45, 52, 54, 56, 57, 63, 65, 70, 72, 73, 74, 76, 78, 81, 84, 90, 91, 95, 97... (sequence A051913 in the OEIS)

The set of n for which a regular n-gon has no solid construction is the sequence

11, 22, 23, 25, 29, 31, 33, 41, 43, 44, 46, 47, 49, 50, 53, 55, 58, 59, 61, 62, 66, 67, 69, 71, 75, 77, 79, 82, 83, 86, 87, 88, 89, 92, 93, 94, 98, 99, 100... (sequence A048136 in the OEIS)

Like the question with Fermat primes, it is an open question as to whether there are an infinite number of Pierpont primes.

Angle Trisection

What if, together with the straightedge and compass, we had a tool that could (only) trisect an arbitrary angle? Such constructions are solid constructions, but there exist numbers with solid constructions that cannot be constructed using such a tool. For example, we cannot double the cube with such a tool. On the other hand, every regular n-gon that has a solid construction can be constructed using such a tool.

Origami

The mathematical theory of origami is more powerful than compass and straightedge construction. Folds satisfying the Huzita–Hatori axioms can construct exactly the same set of points as the extended constructions using a compass and conic drawing tool. Therefore, origami can also be used to solve cubic equations (and hence quartic equations), and thus solve two of the classical problems.

Markable Rulers

Archimedes, Nicomedes and Apollonius gave constructions involving the use of a markable ruler. This would permit them, for example, to take a line segment, two lines (or circles), and a point; and then draw a line which passes through the given point and intersects both lines, and such that the distance between the points of intersection equals the given segment. This the Greeks called neusis ("inclination", "tendency" or "verging"), because the new line tends to the point. In this expanded scheme, we can trisect an arbitrary angle (see Archimedes' trisection) or extract an arbitrary cube root (due to Nicomedes). Hence, any distance whose ratio to an existing distance is the solution of a cubic or a quartic equation is constructible. Regular polygons with solid constructions, like the heptagon, are constructible; and John H. Conway and Richard K. Guy give constructions for several of them;. The neusis construction is more powerful than a conic drawing tool, as one can construct complex numbers that do not have solid constructions. In fact, using this tool one can solve some quintics that are not solvable using radicals. It is known that one cannot solve an irreducible polynomial of prime degree greater or equal to 7 using the neusis construction, so it is not possible to construct a regular 23-gon or 29-gon using this tool. Benjamin and Snyder proved that it is possible to construct the regular 11-gon, but did not give a construction. It is still open as to whether a regular 25-gon or 31-gon is constructible using this tool.

Computation of Binary Digits

In 1998 Simon Plouffe gave a ruler and compass algorithm that can be used to compute binary digits of certain numbers. The algorithm involves the repeated doubling of an angle and becomes physically impractical after about 20 binary digits.

ANGLE

In planar geometry, an angle is the figure formed by two rays, called the sides of the angle, sharing a common endpoint, called the vertex of the angle. Angles formed by two rays lie in a plane, but this plane does not have to be a Euclidean plane. Angles are also formed by the intersection of two planes in Euclidean and other spaces. These are called dihedral angles. Angles formed by the intersection of two curves in a plane are defined as the angle determined by the tangent rays at the point of intersection. Similar statements hold in space, for example, the spherical angle formed by two great circles on a sphere is the dihedral angle between the planes determined by the great circles.

Angle is also used to designate the measure of an angle or of a rotation. This measure is the ratio of the length of a circular arc to its radius. In the case of a geometric angle, the arc is centered at the vertex and delimited by the sides. In the case of a rotation, the arc is centered at the center of the rotation and delimited by any other point and its image by the rotation.

The word angle comes from the Latin word angulus, meaning “corner”; cognate words are the Greek γκύλος (ankylos), meaning crooked, curved, and the English word ankle. Both are connected with the Proto-Indo-European root *ank-, meaning to bend or bow.

Euclid defines a plane angle as the inclination to each other, in a plane, of two lines which meet each other, and do not lie straight with respect to each other. According to Proclus an angle must be either a quality or a quantity, or a relationship. The first concept was used by Eudemus, who regarded an angle as a deviation from a straight line; the second by Carpus of Antioch, who regarded it as the interval or space between the intersecting lines; Euclid adopted the third concept, although his definitions of right, acute, and obtuse angles are certainly quantitative.

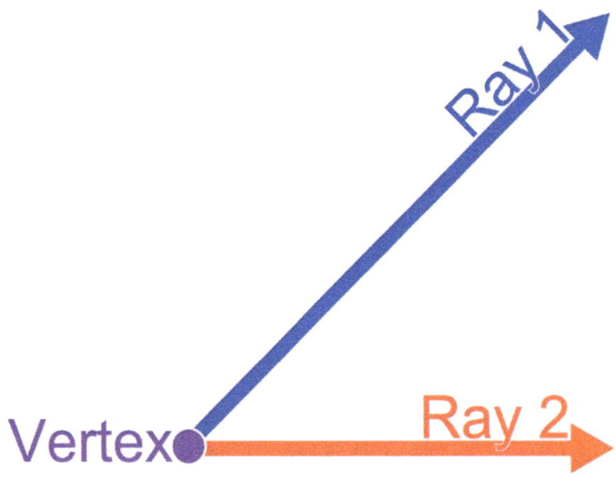

An angle enclosed by rays emanating from a vertex.

∠, the angle symbol in Unicode is U+2220.

Identifying Angles

In mathematical expressions, it is common to use Greek letters (α, β, γ, θ, φ, . . .) to serve as variables standing for the size of some angle. (To avoid confusion with its other meaning, the symbol π is typically not used for this purpose.) Lower case Roman letters (a, b, c, . . .) are also used, as are upper case Roman letters in the context of polygons. See the figures in this article for examples.

In geometric figures, angles may also be identified by the labels attached to the three points that define them. For example, the angle at vertex A enclosed by the rays AB and AC (i.e. the lines from point A to point B and point A to point C) is denoted ∠BAC or $\widehat{\mathbf{BAC}}$. Sometimes, where there is no risk of confusion, the angle may be referred to simply by its vertex ("angle A").

Potentially, an angle denoted, say, ∠BAC might refer to any of four angles: the clockwise angle from B to C, the anticlockwise angle from B to C, the clockwise angle from C to B, or the anticlockwise angle from C to B, where the direction in which the angle is measured determines its sign (see Positive and negative angles). However, in many geometrical situations it is obvious from context that the positive angle less than or equal to 180 degrees is meant, and no ambiguity arises. Otherwise, a convention may be adopted so that ∠BAC always refers to the anticlockwise (positive) angle from B to C, and ∠CAB to the anticlockwise (positive) angle from C to B.

Types of Angles

"Oblique angle" redirects here. For the cinematographic technique, see Dutch angle.

Individual Angles

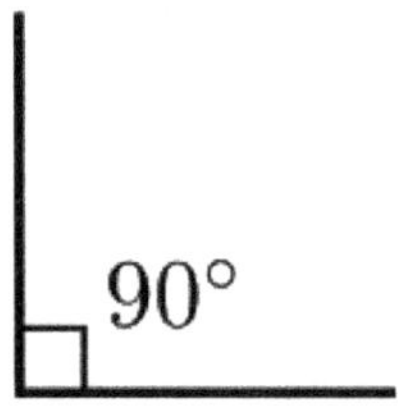

Right angle.

Reflex angle.

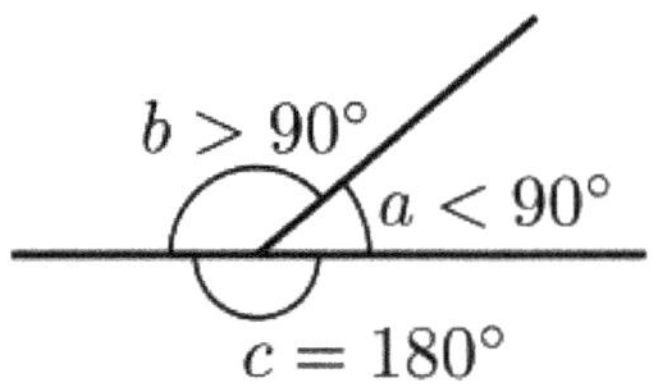

Acute (a), obtuse (b), and straight (c) angles. The acute and obtuse angles are also known as oblique angles.

- Angles smaller than a right angle (less than 90°) are called acute angles ("acute" meaning "sharp").
- An angle equal to 1/4 turn (90° or $\pi/2$ radians) is called a right angle. Two lines that form a right angle are said to be normal, orthogonal, or perpendicular.
- Angles larger than a right angle and smaller than a straight angle (between 90° and 180°) are called obtuse angles ("obtuse" meaning "blunt").

- An angle equal to 1/2 turn (180° or π radians) is called a straight angle.
- Angles larger than a straight angle but less than 1 turn (between 180° and 360°) are called reflex angles.
- An angle equal to 1 turn (360° or 2π radians) is called a full angle, complete angle, or a perigon.
- Angles that are not right angles or a multiple of a right angle are called oblique angles.

The names, intervals, and measured units are shown in a table below:

Name	acute	right angle	obtuse	straight	reflex	perigon
Units	**Interval**					
Turns	(0, 1/4)	1/4	(1/4, 1/2)	1/2	(1/2, 1)	1
Radians	(0, 1/2π)	1/2π	(1/2π, π)	π	(π, 2π)	2π
Degrees	(0, 90)°	90°	(90, 180)°	180°	(180, 360)°	360°
Gons	(0, 100)^g	100^g	(100, 200)^g	200^g	(200, 400)^g	400^g

Equivalence Angle Pairs

- Angles that have the same measure (i.e. the same magnitude) are said to be equal or congruent. An angle is defined by its measure and is not dependent upon the lengths of the sides of the angle (e.g. all right angles are equal in measure).
- Two angles which share terminal sides, but differ in size by an integer multiple of a turn, are called coterminal angles.
- A reference angle is the acute version of any angle determined by repeatedly subtracting or adding straight angle (1/2 turn, 180°, or π radians), to the results as necessary, until the magnitude of result is an acute angle, a value between 0 and 1/4 turn, 90°, or $\pi/2$ radians. For example, an angle of 30 degrees has a reference angle of 30 degrees, and an angle of 150 degrees also has a reference angle of 30 degrees (180 – 150). An angle of 750 degrees has a reference angle of 30 degrees (750 – 720).

Vertical and Adjacent Angle Pairs

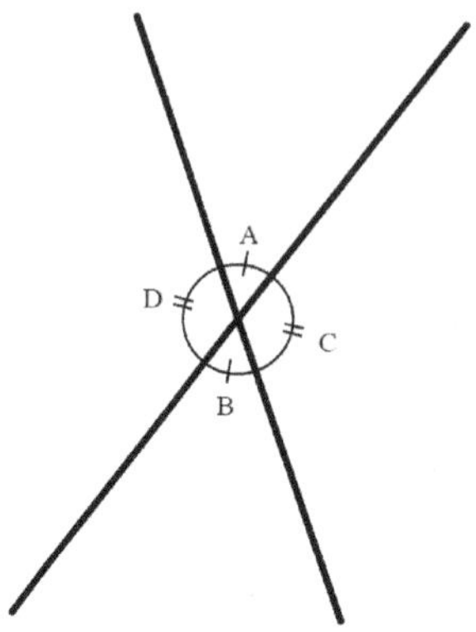

Angles A and B are a pair of vertical angles; angles C and D are a pair of vertical angles.

When two straight lines intersect at a point, four angles are formed. Pairwise these angles are named according to their location relative to each other.

- A pair of angles opposite each other, formed by two intersecting straight lines that form an "X"-like shape, are called vertical angles or opposite angles or vertically opposite angles. They are abbreviated as vert. opp. ∠s.

The equality of vertically opposite angles is called the vertical angle theorem. Eudemus of Rhodes attributed the proof to Thales of Miletus. The proposition showed that since both of a pair of vertical angles are supplementary to both of the adjacent angles, the vertical angles are equal in measure. According to a historical Note, when Thales visited Egypt, he observed that whenever the Egyptians drew two intersecting lines, they would measure the vertical angles to make sure that they were equal. Thales concluded that one could prove that all vertical angles are equal if one accepted some general notions such as: all straight angles are equal, equals added to equals are equal, and equals subtracted from equals are equal.

In the figure, assume the measure of Angle A = x. When two adjacent angles form a straight line, they are supplementary. Therefore, the measure of Angle C = 180 − x. Similarly, the measure of Angle D = 180 − x. Both Angle C and Angle D have measures equal to 180 − x and are congruent. Since Angle B is supplementary to both Angles C and D, either of these angle measures may be used to determine the measure of Angle B. Using the measure of either Angle C or Angle D we find the measure of Angle B = 180 − (180 − x) = 180 − 180 + x = x. Therefore, both Angle A and Angle B have measures equal to x and are equal in measure.

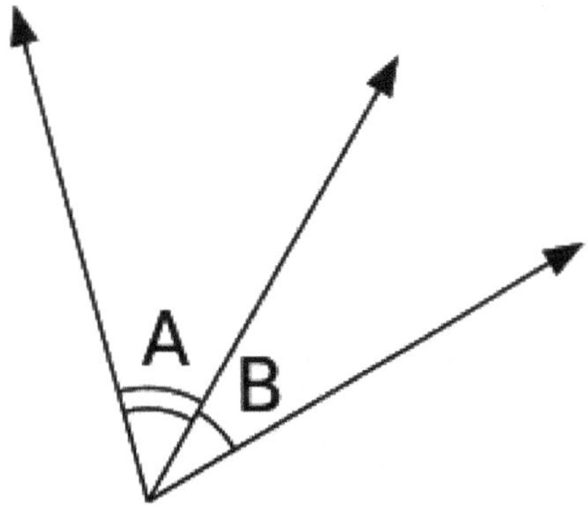

Angles A and B are adjacent.

- Adjacent angles, often abbreviated as adj. ∠s, are angles that share a common vertex and edge but do not share any interior points. In other words, they are angles that are side by side, or adjacent, sharing an "arm". Adjacent angles which sum to a right angle, straight angle or full angle are special and are respectively called complementary, supplementary and explementary angles (see "Combine angle pairs" below).

A transversal is a line that intersects a pair of (often parallel) lines and is associated with alternate interior angles, corresponding angles, interior angles, and exterior angles.

Combining Angle Pairs

There are three special angle pairs which involve the summation of angles:

- **Complementary angles** are angle pairs whose measures sum to one right angle (1/4 turn, 90°, or π/2 radians). If the two complementary angles are adjacent their non-shared sides form a right angle. In Euclidean geometry, the two acute angles in a right triangle are complementary, because the sum of internal angles of a triangle is 180 degrees, and the right angle itself accounts for ninety degrees.

The adjective complementary is from Latin complementum, associated with the verb complere, "to fill up". An acute angle is "filled up" by its complement to form a right angle.

The difference between an angle and a right angle is termed the complement of the angle.

If angles A and B are complementary, the following relationships hold:

$$\sin^2 A + \sin^2 B = 1. \quad \cos^2 A + \cos^2 B = 1.$$
$$\tan A = \cot B. \quad \sec A = \csc B.$$

(The tangent of an angle equals the cotangent of its complement and its secant equals the cosecant of its complement.)

The prefix "co-" in the names of some trigonometric ratios refers to the word "complementary".

- Two angles that sum to a straight angle (1/2 turn, 180°, or π radians) are called supplementary angles.

If the two supplementary angles are adjacent (i.e. have a common vertex and share just one side), their non-shared sides form a straight line. Such angles are called a linear pair of angles. However, supplementary angles do not have to be on the same line, and can be separated in space. For example, adjacent angles of a parallelogram are supplementary, and opposite angles of a cyclic quadrilateral (one whose vertices all fall on a single circle) are supplementary.

If a point P is exterior to a circle with center O, and if the tangent lines from P touch the circle at points T and Q, then ∠TPQ and ∠TOQ are supplementary.

The sines of supplementary angles are equal. Their cosines and tangents (unless undefined) are equal in magnitude but have opposite signs.

In Euclidean geometry, any sum of two angles in a triangle is supplementary to the third, because the sum of internal angles of a triangle is a straight angle.

- Two angles that sum to a complete angle (1 turn, 360°, or 2π radians) are called **explementary angles** or **conjugate angles**.

The difference between an angle and a complete angle is termed the explement of the angle or conjugate of an angle.

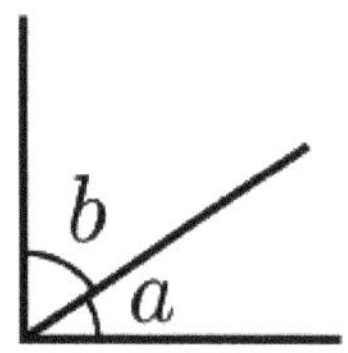

The complementary angles a and b (b is the complement of a, and a is the complement of b).

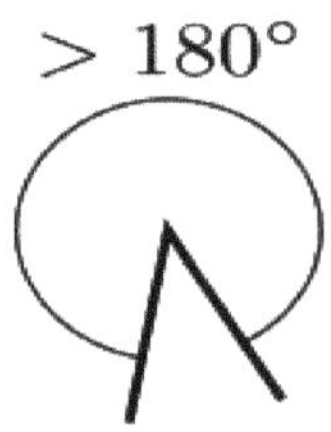

A reflex angle and its conjugate are explementary angles, and their sum is a complete angle.

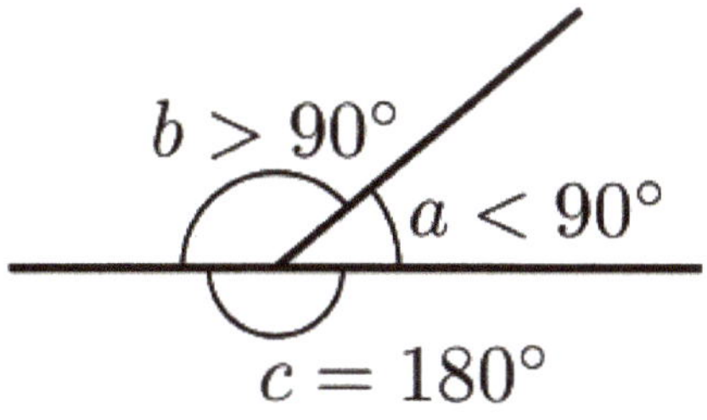

The angles a and b are supplementary angles.

Polygon Related Angles

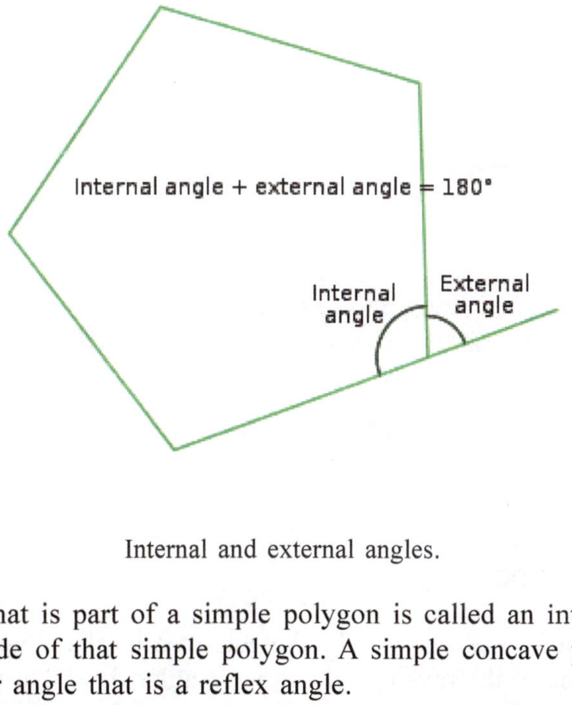

Internal and external angles.

- An angle that is part of a simple polygon is called an interior angle if it lies on the inside of that simple polygon. A simple concave polygon has at least one interior angle that is a reflex angle.

In Euclidean geometry, the measures of the interior angles of a triangle add up to πradians, 180°, or 1/2 turn; the measures of the interior angles of a simple convexquadrilateral add up to 2π radians, 360°, or 1 turn. In general, the measures of the interior angles of a simple convex polygon with n sides add up to (n − 2)π radians, or 180(n − 2) degrees, (2n − 4) right angles, or (n/2 − 1) turn.

- The supplement of an interior angle is called an exterior angle, that is, an interior angle and an exterior angle form a linear pair of angles. There are two exterior angles at each vertex of the polygon, each determined by extending one of the two sides of the polygon that meet at the vertex; these two angles are vertical angles and hence are equal. An exterior angle measures the amount of rotation one has to make at a vertex to trace out the polygon. If the corresponding interior angle is a reflex angle, the exterior angle should be considered negative. Even in a non-simple polygon it may be possible to define the exterior angle, but one will have to pick an orientation of the plane (or surface) to decide the sign of the exterior angle measure.

In Euclidean geometry, the sum of the exterior angles of a simple convex polygon will be one full turn (360°). The exterior angle here could be called a supplementary exterior angle. Exterior angles are commonly used in Logo Turtle Geometry when drawing regular polygons.

- In a triangle, the bisectors of two exterior angles and the bisector of the other interior angle are concurrent (meet at a single point).
- In a triangle, three intersection points, each of an external angle bisector with the opposite extended side, are collinear.
- In a triangle, three intersection points, two of them between an interior angle bisector and the opposite side, and the third between the other exterior angle bisector and the opposite side extended, are collinear.
- Some authors use the name exterior angle of a simple polygon to simply mean the explement exterior angle (not supplement!) of the interior angle. This conflicts with the above usage.

Plane Related Angles

- The angle between two planes (such as two adjacent faces of a polyhedron) is called a dihedral angle. It may be defined as the acute angle between two lines normal to the planes.
- The angle between a plane and an intersecting straight line is equal to ninety degrees minus the angle between the intersecting line and the line that goes through the point of intersection and is normal to the plane.

Measuring Angles

The size of a geometric angle is usually characterized by the magnitude of the smallest rotation that maps one of the rays into the other. Angles that have the same size are said to be equal or congruent or equal in measure.

In some contexts, such as identifying a point on a circle or describing the orientation of an object in two dimensions relative to a reference orientation, angles that differ by an exact multiple of a full turn are effectively equivalent. In other contexts, such as identifying a point on a spiral curve or describing the cumulative rotation of an object in two dimensions relative to a reference orientation, angles that differ by a non-zero multiple of a full turn are not equivalent.

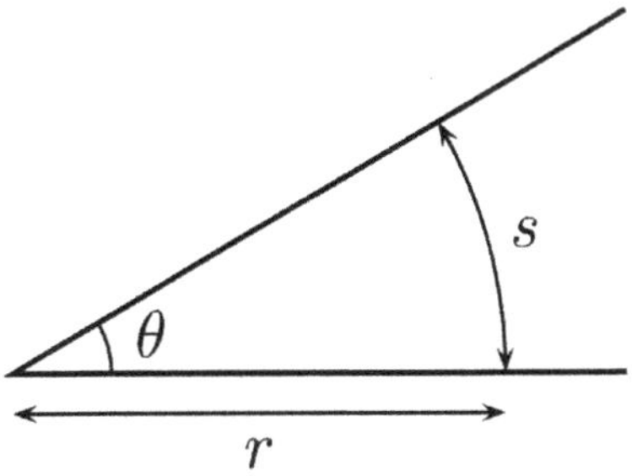

The measure of angle θ (in radians) is the quotient of s and r.

In order to measure an angle θ, a circular arc centered at the vertex of the angle is drawn, e.g. with a pair of compasses. The ratio of the length s of the arc by the radius r of the circle is the measure of the angle in radians.

The measure of the angle in another angular unit is then obtained by multiplying its measure in radians by the scaling factor k/2π, where k is the measure of a complete turn in the chosen unit (for example 360 for degrees or 400 for gradians):

$$\theta = k\frac{s}{2\pi r}.$$

The value of θ thus defined is independent of the size of the circle: if the length of the radius is changed then the arc length changes in the same proportion, so the ratio s/r is unaltered. (Proof. The formula above can be rewritten as k = θr/s. One turn, for which θ = n units, corresponds to an arc equal in length to the circle's circumference, which is 2πr, so s = 2πr. Substituting n for θ and 2πr for s in the formula, results in k = nr/2πr = n/2π.)

Angle Addition Postulate

The angle addition postulate states that if B is in the interior of angle AOC, then $m\angle AOC = m\angle AOB + m\angle BOC$

The measure of the angle AOC is the sum of the measure of angle AOB and the measure of angle BOC. In this postulate it does not matter in which unit the angle is measured as long as each angle is measured in the same unit.

Units

Units used to represent angles are listed below in descending magnitude order. Of these units, the degree and the radian are by far the most commonly used. Angles expressed in radians are dimensionless for the purposes of dimensional analysis.

Most units of angular measurement are defined such that one turn (i.e. one full circle) is equal to n units, for some whole number n. The two exceptions are the radian and the diameter part.

Turn (n = 1)

The turn, also cycle, full circle, revolution, and rotation, is complete circular movement or measure (as to return to the same point) with circle or ellipse. A turn is abbreviated τ, cyc, rev, or rot depending on the application, but in the acronym rpm (revolutions per minute), just r is used. A turn of n units is obtained by setting $k = 1/2\pi$ in the formula above. The equivalence of 1 turn is 360°, 2π rad, 400 grad, and 4 right angles. The symbol τ can also be used as a mathematical constant to represent 2π radians. Used in this way ($k = \tau/2\pi$) allows for radians to be expressed as a fraction of a turn. For example, half a turn is $\tau/2 = \pi$.

Quadrant (n = 4)

The quadrant is 1/4 of a turn, i.e. a right angle. It is the unit used in Euclid's Elements. 1 quad. = 90° = $\pi/2$ rad = 1/4 turn = 100 grad. In German the symbol ∟ has been used to denote a quadrant.

Sextant (n = 6)

The sextant (angle of the equilateral triangle) is 1/6 of a turn. It was the unit used by the Babylonians, and is especially easy to construct with ruler and compasses. The degree, minute of arc and second of arc are sexagesimal subunits of the Babylonian unit. 1 Babylonian unit = 60° = $\pi/3$ rad ≈ 1.047197551 rad.

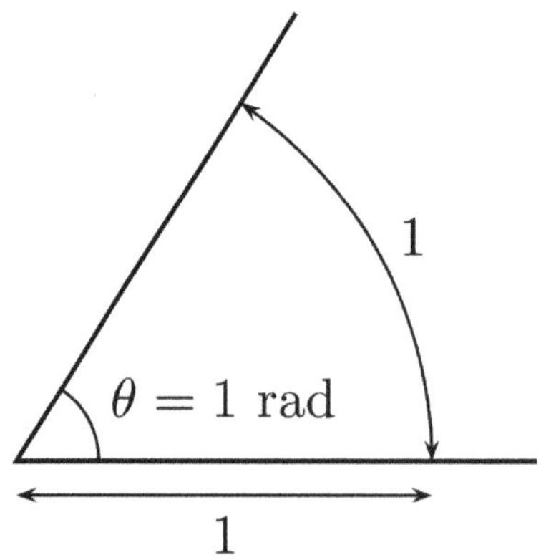

$\theta = s/r$ rad = 1 rad.

Radian ($n = 2\pi = 6.283\ldots$)

The radian is the angle subtended by an arc of a circle that has the same length as the circle's radius. The case of radian for the formula given earlier, a radian of $n = 2\pi$ units is obtained by setting $k = 2\pi/2\pi = 1$. One turn is 2π radians, and one radian is $180/\pi$ degrees, or about 57.2958 degrees. The radian is abbreviated rad, though this symbol is often omitted in mathematical texts, where radians are assumed unless specified otherwise. When radians are used angles are considered as dimensionless. The radian is used in virtually all mathematical work beyond simple practical geometry, due, for example, to the pleasing and "natural" properties that the trigonometric functions display when their arguments are in radians. The radian is the (derived) unit of angular measurement in the SI system.

Clock position (n = 12)

A clock position is the relative direction of an object described using the analogy of a 12-hour clock. One imagines a clock face lying either upright or flat in front of oneself, and identifies the twelve hour markings with the directions in which they point.

Hour angle (n = 24)

The astronomical hour angle is 1/24 of a turn. Since this system is amenable to measuring objects that cycle once per day (such as the relative position of stars), the sexagesimal subunits are called minute of time and second of time. Note that these are distinct from, and 15 times larger than, minutes and seconds of arc. 1 hour = 15° = $\pi/12$ rad = 1/6 quad. = 1/24 turn = 16 2/3 grad.

(Compass) point or wind (n = 32)

The point, used in navigation, is 1/32 of a turn. 1 point = 1/8 of a right angle = 11.25° = 12.5 grad. Each point is subdivided in four quarter-points so that 1 turn equals 128 quarter-points.

Hexacontade (n = 60)

The hexacontade is a unit of 6° that Eratosthenes used, so that a whole turn was divided into 60 units.

Pechus (n = 144–180)

–The pechus was a Babylonian unit equal to about 2° or 2 1/2°.

Binary degree (n = 256)

The binary degree, also known as the binary radian (or brad), is 1/256 of a turn. The binary degree is used in computing so that an angle can be efficiently represented in a single byte (albeit to limited precision). Other measures of angle used in computing may be based on dividing one whole turn into 2^n equal parts for other values of n.

Degree (n = 360)

The degree, denoted by a small superscript circle (°), is 1/360 of a turn, so one turn is 360°. The case of degrees for the formula given earlier, a degree of n = 360° units is obtained by setting k = 360°/2π. One advantage of this old sexagesimal subunit is that many angles common in simple geometry are measured as a whole number of degrees. Fractions of a degree may be written in normal decimal notation (e.g. 3.5° for three and a half degrees), but the "minute" and "second" sexagesimal subunits of the degree-minute-second system are also in use, especially for geographical coordinates and in astronomy and ballistics:

Diameter part (n = 376.99 . . .)

The diameter part (occasionally used in Islamic mathematics) is 1/60 radian. One "diameter part" is approximately 0.95493°. There are about 376.991 diameter parts per turn.

Grad (*n* = 400)

The grad, also called grade, gradian, or gon, is 1/400 of a turn, so a right angle is 100 grads. It is a decimal subunit of the quadrant. A kilometre was historically defined as a centi-grad of arc along a great circle of the Earth, so the kilometer is the decimal analog to the sexagesimal nautical mile. The grad is used mostly in triangulation.

Mil (*n* = 6000–6400)

The mil is any of several units that are approximately equal to a milliradian. There are several definitions ranging from 0.05625 to 0.06 degrees (3.375 to 3.6 minutes), with the milliradian being approximately 0.05729578 degrees (3.43775 minutes). In NATO countries, it is defined as 1/6400 of a circle. Its value is approximately equal to the angle subtended by a width of 1 metre as seen from 1 km away (2π/6400 = 0.0009817 . . . ≈ 1/1000).

Minute of arc (n = 21,600)

The minute of arc (or MOA, arcminute, or just minute) is 1/60 of a degree = 1/21,600 turn. It is denoted by a single prime (′). For example, 3° 30′ is equal to 3 × 60 + 30 = 210 minutes or 3 + 30/60 = 3.5 degrees. A mixed format with decimal fractions is also sometimes used, e.g. 3° 5.72′ = 3 + 5.72/60 degrees. A nautical mile was historically defined as a minute of arc along a great circle of the Earth.

Second of arc (*n* = 1,296,000)

The second of arc (or arcsecond, or just second) is 1/60 of a minute of arc and 1/3600 of a degree. It is denoted by a double prime (″). For example, 3° 7′ 30″ is equal to 3 + 7/60 + 30/3600 degrees, or 3.125 degrees.

Positive and Negative Angles

Although the definition of the measurement of an angle does not support the concept of a negative angle, it is frequently useful to impose a convention that allows positive and negative angular values to represent orientations and/or rotations in opposite directions relative to some reference.

In a two-dimensional Cartesian coordinate system, an angle is typically defined by its two sides, with its vertex at the origin. The initial side is on the positive x-axis, while the other side or terminal side is defined by the measure from the initial side in radians, degrees, or turns. With positive angles representing rotations toward the positive y-axis and negative angles representing rotations toward the negative y-axis. When Cartesian coordinates are represented by standard position, defined by the x-axis rightward and the y-axis upward, positive rotations are anticlockwise and negative rotations are clockwise.

In many contexts, an angle of $-\theta$ is effectively equivalent to an angle of one full turn minus θ. For example, an orientation represented as −45° is effectively equivalent to an orientation represented as 360° − 45° or 315°. Although the final position is the same, a physical rotation (movement) of −45° is not the same as a rotation of 315° (for example, the rotation of a person holding a broom resting on a dusty floor would leave visually different traces of swept regions on the floor).

In three-dimensional geometry, clockwise and anticlockwise have no absolute meaning, so the direction of positive and negative angles must be defined relative to some reference, which is typically a vector passing through the angle's vertex and perpendicular to the plane in which the rays of the angle lie.

In navigation, bearings are measured relative to north. By convention, viewed from above, bearing angles are positive clockwise, so a bearing of 45° corresponds to a north-east orientation. Negative bearings are not used in navigation, so a north-west orientation corresponds to a bearing of 315°.

Alternative Ways of Measuring the Size of an Angle

There are several alternatives to measuring the size of an angle by the angle of rotation. The grade of a slope, or gradient is equal to the tangent of the angle, or sometimes (rarely) the sine. A gradient is often expressed as a percentage. For very small values (less than 5%), the grade of a slope is approximately the measure of the angle in radians.

In rational geometry the spread between two lines is defined at the square of the sine of the angle between the lines. Since the sine of an angle and the sine of its supplementary angle are the same, any angle of rotation that maps one of the lines into the other leads to the same value for the spread between the lines.

Astronomical Approximations

Astronomers measure angular separation of objects in degrees from their point of observation.

- 0.5° is approximately the width of the sun or moon.
- 1° is approximately the width of a little finger at arm's length.
- 10° is approximately the width of a closed fist at arm's length.
- 20° is approximately the width of a handspan at arm's length.

These measurements clearly depend on the individual subject, and the above should be treated as rough rule of thumb approximations only.

Angles Between Curves

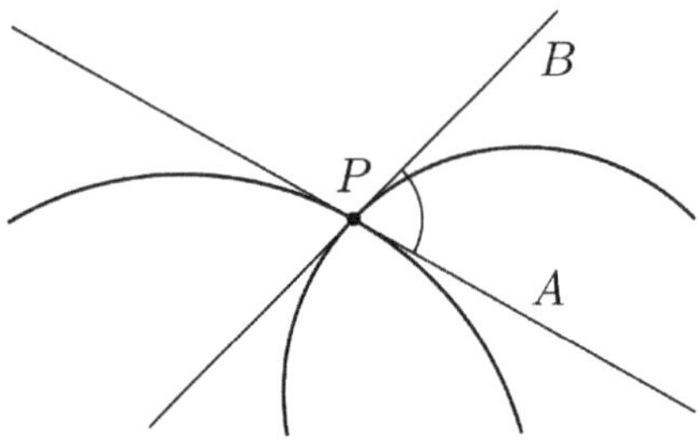

The angle between the two curves at P is defined as the angle between the tangents A and B at P.

The angle between a line and a curve (mixed angle) or between two intersecting curves (curvilinear angle) is defined to be the angle between the tangents at the point of intersection. Various names (now rarely, if ever, used) have been given to particular cases:—amphicyrtic (Gr. μφί, on both sides, κυρτός, convex) or cissoidal (Gr. κισσός, ivy), biconvex; xystroidal or sistroidal (Gr. ξυστρίς, a tool for scraping), concavo-convex; amphicoelic (Gr. κοίλη, a hollow) or angulus lunularis, biconcave.

Bisecting and Trisecting Angles

The ancient Greek mathematicians knew how to bisect an angle (divide it into two angles of equal measure) using only a compass and straightedge, but could only trisect certain angles. In 1837 Pierre Wantzel showed that for most angles this construction cannot be performed.

Dot Product and Generalisations

In the Euclidean space, the angle θ between two Euclidean vectors **u** and **v** is related to their dot product and their lengths by the formula

$$\mathbf{u} \cdot \mathbf{v} = \cos(\theta) \left\|\mathbf{u}\right\| \left\|\mathbf{v}\right\|.$$

This formula supplies an easy method to find the angle between two planes (or curved surfaces) from their normal vectors and between skew lines from their vector equations.

Inner Product

To define angles in an abstract real inner product space, we replace the Euclidean dot product (·) by the inner product $\langle\cdot,\cdot\rangle$, i.e

$$\langle \mathbf{u}, \mathbf{v} \rangle = \cos(\theta) \left\|\mathbf{u}\right\| \left\|\mathbf{v}\right\|.$$

In a complex inner product space, the expression for the cosine above may give non-real values, so it is replaced with

$$\operatorname{Re}(\langle \mathbf{u}, \mathbf{v} \rangle) = \cos(\theta) \left\|\mathbf{u}\right\| \left\|\mathbf{v}\right\|.$$

or, more commonly, using the absolute value, with

$$|\langle \mathbf{u}, \mathbf{v} \rangle| = |\cos(\theta)| \left\|\mathbf{u}\right\| \left\|\mathbf{v}\right\|.$$

The latter definition ignores the direction of the vectors and thus describes the angle between one-dimensional subspaces $\operatorname{span}(\mathbf{u})$ and $\operatorname{span}(\mathbf{v})$ spanned by the vectors $\mathbf{u}$ and $\mathbf{v}$ correspondingly.

Angles between Subspaces

The definition of the angle between one-dimensional subspaces $\operatorname{span}(\mathbf{u})$ and $\operatorname{span}(\mathbf{v})$ given by

$$|\langle \mathbf{u}, \mathbf{v} \rangle| = |\cos(\theta)| \left\|\mathbf{u}\right\| \left\|\mathbf{v}\right\|$$

in a Hilbert space can be extended to subspaces of any finite dimensions. Given two subspaces $\mathcal{U}, \mathcal{W}$ with $\dim(\mathcal{U}) := k \leq \dim(\mathcal{W}) := l$, this leads to a definition of k angles called canonical or principal angles between subspaces.

Angles in Riemannian Geometry

In Riemannian geometry, the metric tensor is used to define the angle between two tangents. Where U and V are tangent vectors and g_{ij} are the components of the metric tensor G,

$$\cos\theta = \frac{g_{ij}U^iV^j}{\sqrt{\left|g_{ij}U^iU^j\right|\left|g_{ij}V^iV^j\right|}}.$$

Angles in Geography and Astronomy

In geography, the location of any point on the Earth can be identified using a geographic coordinate system. This system specifies the latitude and longitude of any location in terms of angles subtended at the centre of the Earth, using the equator and (usually) the Greenwich meridian as references.

In astronomy, a given point on the celestial sphere (that is, the apparent position of an astronomical object) can be identified using any of several astronomical coordinate systems, where the references vary according to the particular system. Astronomers measure the angular separation of two stars by imagining two lines through the centre of the Earth, each intersecting one of the stars. The angle between those lines can be measured, and is the angular separation between the two stars.

In both geography and astronomy, a sighting direction can be specified in terms of a vertical angle such as altitude /elevation with respect to the horizon as well as the azimuth with respect to north.

Astronomers also measure the apparent size of objects as an angular diameter. For example, the full moon has an angular diameter of approximately 0.5°, when viewed from Earth. One could say, "The Moon's diameter subtends an angle of half a degree." The small-angle formula can be used to convert such an angular measurement into a distance/size ratio.

CURVE

In mathematics, a **curve** (also called a **curved line** in older texts) is, generally speaking, an object similar to a line but that need not be straight. Thus, a curve is a generalization of a line, in that curvature is not necessarily zero.

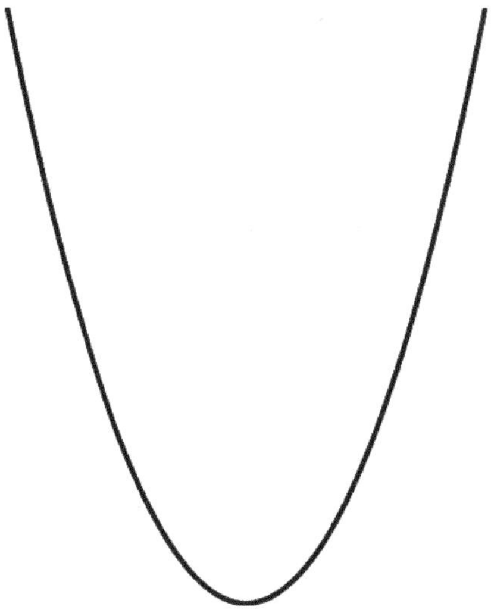

A parabola, a simple example of a curve

Various disciplines within mathematics have given the term different meanings depending on the area of study, so the precise meaning depends on context. However, many of these meanings are special instances of the definition which follows. A curve is a topological space which is locally homeomorphic to a line. In everyday language, this means that a curve is a set of points which, near each of its points, looks like a line, up to a deformation. A simple example of a curve is the parabola, shown to the right. A large number of other curves have been studied in multiple mathematical fields.

A **closed curve** is a curve that forms a path whose starting point is also its ending point—that is, a path from any of its points to the same point.

Closely related meanings include the graph of a function (as in Phillips curve) and a two-dimensional graph.

Megalithic art from Newgrange showing an early interest in curves

Interest in curves began long before they were the subject of mathematical study. This can be seen in numerous examples of their decorative use in art and on everyday objects dating back to prehistoric times. Curves, or at least their graphical representations, are simple to create, for example by a stick in the sand on a beach.

Historically, the term "line" was used in place of the more modern term "curve". Hence the phrases "straight line" and "right line" were used to distinguish what are today called lines from "curved lines". For example, in Book I of Euclid's Elements, a line is defined as a "breadthless length" (Def. 2), while a straight line is defined as "a line that lies evenly with the points on itself" (Def. 4). Euclid's idea of a line is perhaps clarified by the statement "The extremities of a line are points," (Def. 3). Later commentators further classified lines according to various schemes. For example:

- Composite lines (lines forming an angle)
- Incomposite lines
    - Determinate (lines that do not extend indefinitely, such as the circle)
    - Indeterminate (lines that extend indefinitely, such as the straight line and the parabola)

The curves created by slicing a cone (conic sections) were among the curves studied in ancient Greece.

The Greek geometers had studied many other kinds of curves. One reason was their interest in solving geometrical problems that could not be solved using standard compass and straightedge construction. These curves include:

- The conic sections, deeply studied by Apollonius of Perga
- The cissoid of Diocles, studied by Diocles and used as a method to double the cube.
- The conchoid of Nicomedes, studied by Nicomedes as a method to both double the cube and to trisect an angle.
- The Archimedean spiral, studied by Archimedes as a method to trisect an angle and square the circle.
- The spiric sections, sections of tori studied by Perseus as sections of cones had been studied by Apollonius.

A fundamental advance in the theory of curves was the advent of analytic geometry in the seventeenth century. This enabled a curve to be described using an equation rather than an elaborate geometrical construction. This not only allowed new curves to be defined and studied, but it enabled a formal distinction to be made between curves that can be defined using algebraic equations, algebraic curves, and those that cannot, transcendental curves. Previously, curves had been described as "geometrical" or "mechanical" according to how they were, or supposedly could be, generated.

Conic sections were applied in astronomy by Kepler. Newton also worked on an early example in the calculus of variations. Solutions to variational problems, such as the brachistochrone and tautochrone questions, introduced properties of curves in new ways (in this case, the cycloid). The catenary gets its name as the solution to the problem of a hanging chain, the sort of question that became routinely accessible by means of differential calculus.

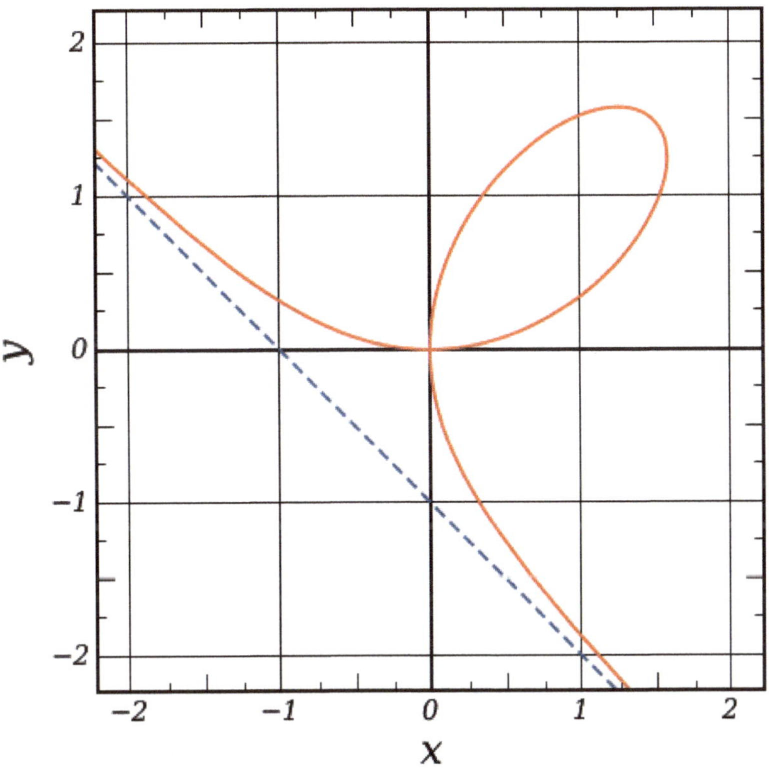

Analytic geometry allowed curves, such as the Folium of Descartes, to be defined using equations instead of geometrical construction.

In the eighteenth century came the beginnings of the theory of plane algebraic curves, in general. Newton had studied the cubic curves, in the general description of the real points into ‹ovals'. The statement of Bézout's theorem showed a number of aspects which were not directly accessible to the geometry of the time, to do with singular points and complex solutions.

From the nineteenth century there is not a separate curve theory, but rather the appearance of curves as the one-dimensional aspect of projective geometry, and differential geometry; and later topology, when for example the Jordan curve theorem was understood to lie quite deep, as well as being required in complex analysis. The era of the space-filling curves finally provoked the modern definitions of curve.

Definition

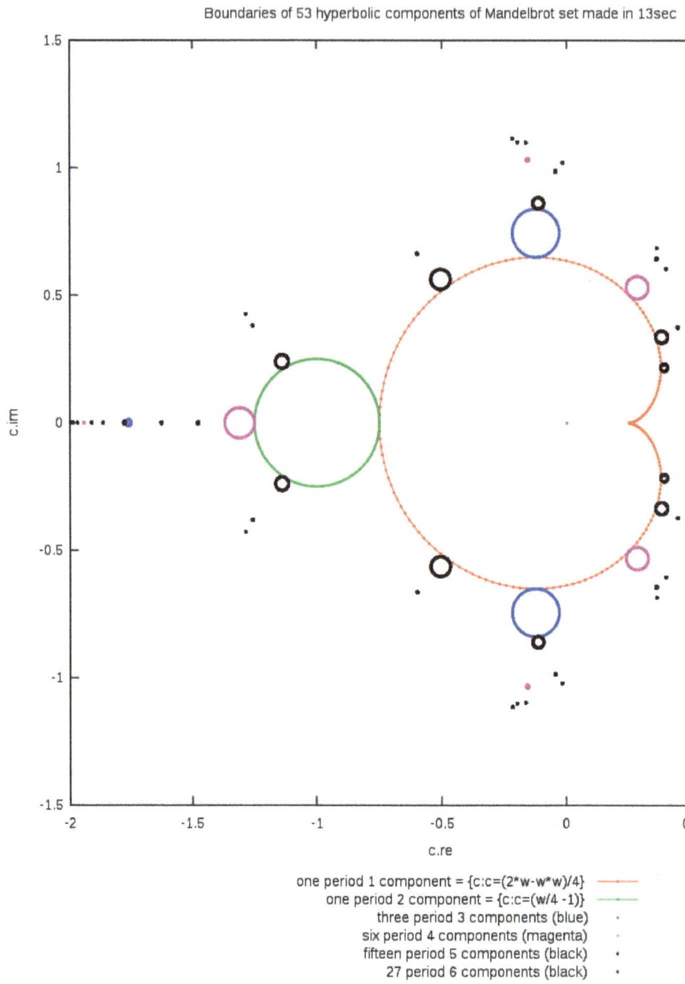

Boundaries of hyperbolic components of Mandelbrot set as closed curves

In general, a **curve** is defined through a continuous function $\gamma: I \to X$ from an interval I of the real numbers into a topological space X. Depending on the context, it is either γ or its image $\gamma(I)$ which is called a curve.

In general topology, when non-differentiable functions are considered, it is the map γ, which is called a curve, because its image may look very differently from what is commonly called a curve. For example, the image of the Peano curve completely fills the square. On the other hand, when one considers curves defined by a differentiable function (or, at least, a piecewise differentiable function), this is commonly the image of the function which is called a curve.

- The curve is said to be **simple**, or a **Jordan arc**, if γ is injective, i.e. if for all x, y in I, we have $\gamma(x) = \gamma(y)$ implies $x = y$. If I is a closed bounded interval $[a, b]$, we also allow the possibility $\gamma(a) = \gamma(b)$ (this convention makes it possible to talk about "closed" simple curves, see below). In other words, this curve "does not cross itself and has no missing points".

- If $\gamma(x) = \gamma(y)$ for some $x \neq y$ (other than the extremities of I) then $\gamma(x)$ is called a **double** (or **multiple**) **point** of the curve. This is a special case of a singular point of a curve.
- A curve γ, is said to be **closed** or **a loop** if $I = [a, b]$ and if $\gamma(a) = \gamma(b)$. A closed curve is thus the image of a continuous mapping of the circle S^1 a **simple closed curve** is also called a **Jordan curve**. The Jordan curve theorem states that such curves divide the plane into an "interior" and an "exterior".

A **plane curve** is a curve for which X is the Euclidean plane—these are the examples first encountered—or in some cases the projective plane. A **space curve** is a curve for which X is of three dimensions, usually Euclidean space; a **skew curve** is a space curve which lies in no plane. These definitions of plane, space and skew curves apply also to real algebraic curves, although the above definition of a curve does not applies (a real algebraic curve may be disconnected).

This definition of curve captures our intuitive notion of a curve as a connected, continuous geometric figure that is like a line, without thickness and drawn without interruption, although it also includes figures that can hardly be called curves in common usage. For example, the image of a curve can cover a square in the plane (space-filling curve). The image of simple plane curve can have Hausdorff dimension bigger than one (see Koch snowflake) and even positive Lebesgue measure (the last example can be obtained by small variation of the Peano curve construction). The dragon curve is another unusual example.

Differentiable Curve

Roughly speaking a **differentiable curve** is a curve that is defined as being locally the image of an injective differentiable function $\gamma: I \rightarrow X$ from an interval I of the real numbers into a differentiable manifold X, often $\mathbb{R}^n$.

More precisely, a differentiable curve is a subset C of X where every point of C has a neighborhood U such that $C \cap U$ is diffeomorphic to an interval of the real numbers. In other words, a differentiable curve differentiable manifold is of dimension one.

Length of a Curve

If $X = \mathbb{R}^n$ is an Euclidean space and $\gamma : [a, b] \rightarrow \mathbb{R}^n$ is an injective differentiable function, then the image of γ is a curve of length

$$\text{Length}(\gamma) \stackrel{\text{df}}{=} \int_a^b |\gamma'(t)|\, dt.$$

This length is independent of the choice of the function g that has been chosen for parameterizing the curve.

In particular, the length γ of the graph of a differentiable function y = f(x) defined on the interval [a, b] is

$$s = \int_a^b \sqrt{1 + [f'(x)]^2}\, dx.$$

More generally, if X is a metric space with metric d, then we can define the length of a curve defined by $\gamma : [a, b] \to X$ by

$$\text{Length}(\gamma) \stackrel{\text{df}}{=} \sup\left(\left\{\sum_{i=1}^{n} d(\gamma(t_i), \gamma(t_{i-1})) \,\middle|\, n \in \mathbb{N} \text{ and } a = t_0 < t_1 < \ldots < t_n = b\right\}\right),$$

where the sup is over all n and all partitions $t_0 < t_1 < \ldots < t_n$ of $[a, b]$.

A **rectifiable curve** is a curve with finite length. A parametrization of $\gamma : [a, b] \to X$ is called **natural** (or **unit speed** or **parametrised by arc length**) if for any $t_1, t_2 \in [a, b]$ such that $t_1 \leq t_2$, we have

$$\text{Length}\left(\gamma|_{[t_1,t_2]}\right) = t_2 - t_1.$$

If $\gamma : [a, b] \to X$ is a Lipschitz-continuous function, then it is automatically rectifiable. Moreover, in this case, one can define the **speed** (or metric derivative) of γ at $t \in [a, b]$ as

$$\text{Speed}_\gamma(t) \stackrel{\text{df}}{=} \limsup_{[a,b]\ni s \to t} \frac{d(\gamma(s), \gamma(t))}{|s - t|}$$

and then

$$\text{Length}(\gamma) = \int_a^b \text{Speed}_\gamma(t)\, dt.$$

Differential Geometry

While the first examples of curves that are met are mostly plane curves (that is, in everyday words, curved lines in two-dimensional space), there are obvious examples such as the helix which exist naturally in three dimensions. The needs of geometry, and also for example classical mechanics are to have a notion of curve in space of any number of dimensions. In general relativity, a world line is a curve in spacetime.X is a differentiable manifold, then we can define the notion of differentiable curve in X This general idea is enough to cover many of the applications of curves in mathematics. From a local point of view one can take X to be Euclidean space. On the other hand, it is useful to be more general, in that (for example) it is possible to define the tangent vectors to X by means of this notion of curve.

If X is a smooth manifold, a smooth curve in X is a smooth map

$\gamma: I \to X$.

This is a basic notion. There are less and more restricted ideas, too. If X is a C^k manifold (i.e., a manifold whose charts are k times continuously differentiable), then a C^k curve in X is such a curve which is only assumed to be C^k (i.e.k times continuously differentiable). If X is an analytic manifold (i.e. infinitely differentiable and charts are expressible as power series), and γ is an analytic map, then γ is said to be an analytic curve.

A differentiable curve is said to be regular if its derivative never vanishes. (In words, a regular curve never slows to a stop or backtracks on itself.) Two Ck differentiable curves

$\gamma_1: I \to X$ and

$\gamma_2: J \to X$

are said to be equivalent if there is a bijective C^k map

$p: J \to I$

such that the inverse map

$p^{-1}: I \to J$

is also C^k and

$\gamma_2(t) = \gamma_1(p(t))$

for all t. The map γ_2 is called a reparametrisation of γ_1; and this makes an equivalence relation on the set of all C^k differentiable curves in X. A C^k arc is an equivalence class of C^k curves under the relation of reparametrisation.

Algebraic Curve

Algebraic curves are the curves considered in algebraic geometry. A plane algebraic curve is the locus of the points of coordinates x, y such that f(x, y) = 0, where f is a polynomial in two variables defined over some field F. Algebraic geometry normally looks not only on points with coordinates in F but on all the points with coordinates in an algebraically closed field K. If C is a curve defined by a polynomial f with coefficients in F, the curve is said **defined** over F. The points of the curve C with coordinates in a field G are said **rational** over G and can be denoted C(G)). When G is the field of the rational numbers, one simply talks of rational points. For example, Fermat's Last Theorem may be restated as: For n > 2, every rational point of the Fermat curve of degree n has a zero coordinate.

Algebraic curves can also be space curves, or curves in a space of higher dimension, say n. They are defined as algebraic varieties of dimension one. They may be obtained as the common solutions of at least n–1 polynomial equations in n variables. If n–1 polynomials are sufficient to define a curve in a space of dimension n, the curve is said to be a complete intersection. By eliminating variables (by any tool of elimination theory), an algebraic curve may be projected onto a plane algebraic curve, which however may introduce new singularities such as cusps or double points.

A plane curve may also be completed in a curve in the projective plane: if a curve is defined by a polynomial f of total degree d, then $w^d f(u/w, v/w)$ simplifies to a homogeneous polynomial g(u, v, w) of degree d. The values of u, v, w such that g(u, v, w) = 0 are the homogeneous coordinates of the points of the completion of the curve in the projective plane and the points of the initial curve are those such w is not zero. An example is the Fermat curve $u^n + v^n = w^n$, which has an affine form $x^n + y^n = 1$. A similar process of homogenization may be defined for curves in higher dimensional spaces

Important examples of algebraic curves are the conics, which are nonsingular curves of degree two and genus zero, and elliptic curves, which are nonsingular curves of genus one studied in number theory and which have important applications to cryptography. Because algebraic curves in fields of characteristic zero are most often studied over the complex numbers, algebraic curves in algebraic geometry may be considered as real surfaces. In particular, the non-singular complex projective algebraic curves are called Riemann surfaces.

Diagonal

In geometry, a **diagonal** is a line segment joining two vertices of a polygon or polyhedron, when those vertices are not on the same edge. Informally, any sloping line is called diagonal. The word "diagonal" derives from the ancient Greek διαγώνιος diagonios, "from angle to angle" (from διά- dia-, "through", "across" and γωνία gonia, "angle", related to gony "knee"); it was used by both Strabo and Euclid to refer to a line connecting two vertices of a rhombus or cuboid, and later adopted into Latin as diagonus ("slanting line").

In matrix algebra, a diagonal of a square matrix is a set of entries extending from one corner to the farthest corner.

There are also other, non-mathematical uses.

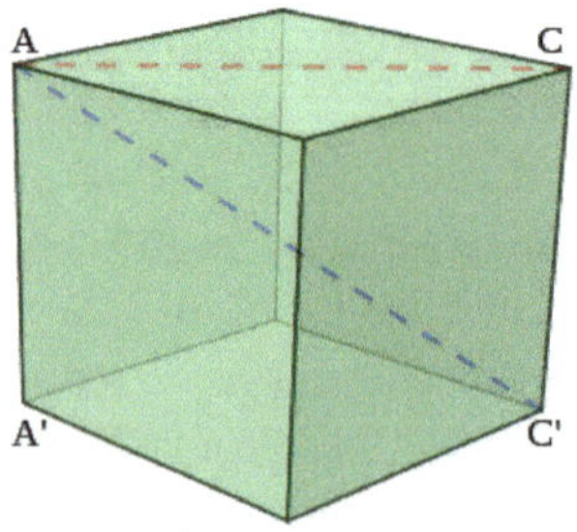

The diagonals of a cube with side length 1. AC' (shown in blue) is a space diagonal with length $\sqrt{3}$, while AC (shown in red) is a face diagonal and has length $\sqrt{2}$.

Non-Mathematical Uses

A stand of basic scaffolding on a house construction site, with diagonal braces to maintain its structure

In engineering, a diagonal brace is a beam used to brace a rectangular structure (such as scaffolding) to withstand strong forces pushing into it; although called a diagonal, due to practical considerations diagonal braces are often not connected to the corners of the rectangle.

Diagonal pliers are wire-cutting pliers defined by the cutting edges of the jaws intersects the joint rivet at an angle or "on a diagonal", hence the name.

A diagonal lashing is a type of lashing used to bind spars or poles together applied so that the lashings cross over the poles at an angle.

In association football, the diagonal system of control is the method referees and assistant referees use to position themselves in one of the four quadrants of the pitch.

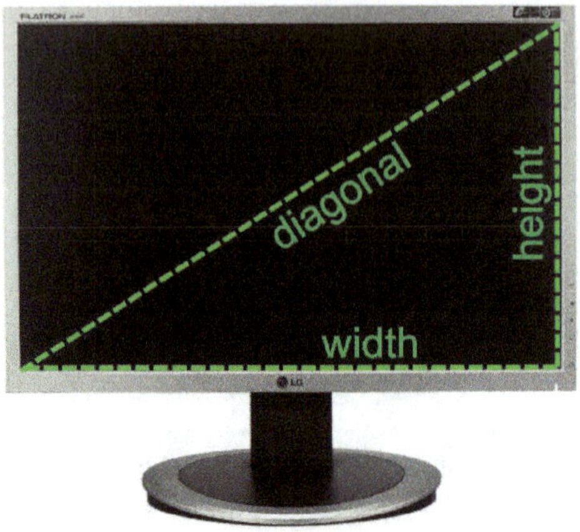

The diagonal is a common measurement of display size.

Polygons

As applied to a polygon, a diagonal is a line segment joining any two non-consecutive vertices. Therefore, a quadrilateral has two diagonals, joining opposite pairs of vertices. For any convex polygon, all the diagonals are inside the polygon, but for re-entrant polygons, some diagonals are outside of the polygon.

Any n-sided polygon (n ≥ 3), convex or concave, has $\frac{n(n-3)}{2}$ diagonals, as each vertex has diagonals to all other vertices except itself and the two adjacent vertices, or n − 3 diagonals, and each diagonal is shared by two vertices.

Sides	Diagonals	Sides	Diagonals	Sides	Diagonals	Sides	Diagonals	Sides	Diagonals
3	0	11	44	19	152	27	324	35	560
4	2	12	54	20	170	28	350	36	594
5	5	13	65	21	189	29	377	37	629
6	9	14	77	22	209	30	405	38	665
7	14	15	90	23	230	31	434	39	702
8	20	16	104	24	252	32	464	40	740
9	27	17	119	25	275	33	495	41	779
10	35	18	135	26	299	34	527	42	819

Regions formed by Diagonals

In a convex polygon, if no three diagonals are concurrent at a single point, the number of regions that the diagonals divide the interior into is given by

$$\binom{n}{4} + \binom{n-1}{2} = \frac{(n-1)(n-2)(n^2-3n+12)}{24}.$$

For n-gons with n=3. 4, ... the number of regions is

1, 4, 11, 25, 50, 91, 154, 246...

This is OEIS sequence A006522.

Matrices

In the case of a square matrix, the main or principal diagonal is the diagonal line of entries running from the top-left corner to the bottom-right corner. For a matrix A with row index specified by i and column index specified by j, these would be entries A_{ij} with i=j. For example, the identity matrix can be defined as having entries of 1 on the main diagonal and zeroes elsewhere:

$$\begin{pmatrix} 1 & 0 & 0 \\ 0 & 1 & 0 \\ 0 & 0 & 1 \end{pmatrix}$$

The top-right to bottom-left diagonal is sometimes described as the minor diagonal or antidiagonal. The off-diagonal entries are those not on the main diagonal. A diagonal matrix is one whose off-diagonal entries are all zero.

A superdiagonal entry is one that is directly above and to the right of the main diagonal. Just as diagonal entries are those A_{ij} with $j = i$, the superdiagonal entries are those with $j = i + 1$. For example, the non-zero entries of the following matrix all lie in the superdiagonal:

$$\begin{pmatrix} 0 & 2 & 0 \\ 0 & 0 & 3 \\ 0 & 0 & 0 \end{pmatrix}$$

Likewise, a subdiagonal entry is one that is directly below and to the left of the main diagonal, that is, an entry A_{ij} with $j = i - 1$. General matrix diagonals can be specified by an index k measured relative to the main diagonal: the main diagonal has k=0; the superdiagonal has k=1; the subdiagonal has k=-1; and in general, the k-diagonal consists of the entries A_{ij} with $j = i + k$

Geometry

By analogy, the subset of the Cartesian product X×X of any set X with itself, consisting of all pairs (x,x), is called the diagonal, and is the graph of the equality relation on X or equivalently the graph of the identity function from X to x. This plays an important part in geometry; for example, the fixed points of a mapping F from X to itself may be obtained by intersecting the graph of F with the diagonal.

In geometric studies, the idea of intersecting the diagonal with itself is common, not directly, but by perturbing it within an equivalence class. This is related at a deep level with the Euler characteristic and the zeros of vector fields. For example, the circle S^1

has Betti numbers 1, 1, 0, 0, 0, and therefore Euler characteristic 0. A geometric way of expressing this is to look at the diagonal on the two-torus S^1xS^1 and observe that it can move off itself by the small motion (θ, θ) to (θ, θ + ε). In general, the intersection number of the graph of a function with the diagonal may be computed using homology via the Lefschetz fixed point theorem; the self-intersection of the diagonal is the special case of the identity function.

ORTHOGONALITY

In mathematics, **orthogonality** is the relation of two lines at right angles to one another (perpendicularity), and the generalization of this relation into n dimensions; and to a variety of mathematical relations thought of as describing non-overlapping, uncorrelated, or independent objects of some kind.

The concept of orthogonality has been broadly generalized in mathematics (including in the areas of mathematical functions, calculus and linear algebra), as well as in areas such as chemistry, and engineering.

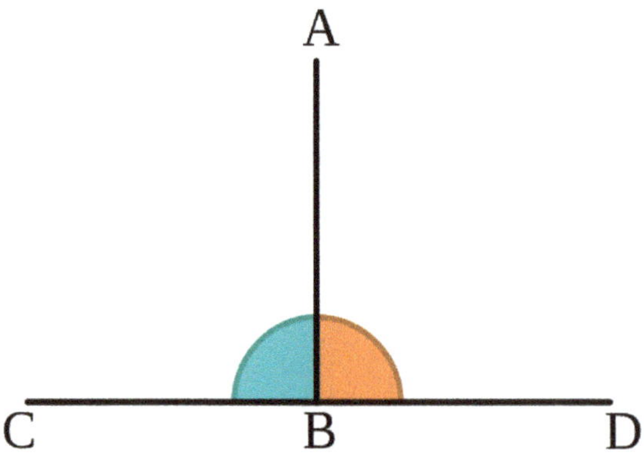

The line segments AB and CD are orthogonal to each other.

Etymology

The word comes from the Greek ὀρθός (orthos), meaning "upright", and γωνία (gonia), meaning "angle". The ancient Greek ρθογώνιον orthogōnion and classical Latin orthogonium originally denoted a rectangle. Later, they came to mean a right triangle. In the 12th century, the post-classical Latin word orthogonalis came to mean a right angle or something related to a right angle.

Mathematics and Physics

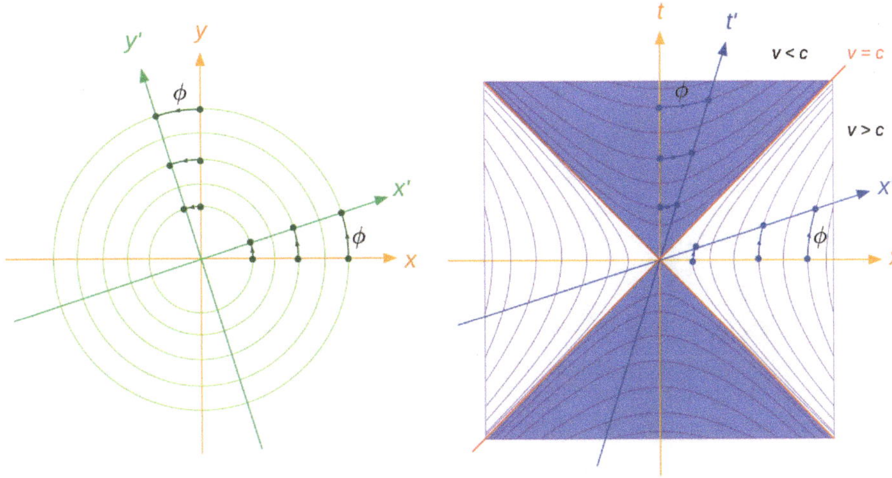

Orthogonality and rotation of coordinate systems compared between **left:** Euclidean space through circular angle ϕ, **right:** in Minkowski spacetime through hyperbolic angle ϕ (red lines labelled c denote the worldlines of a light signal, a vector is orthogonal to itself if it lies on this line).

Definitions

- In geometry, two Euclidean vectors are **orthogonal** if they are perpendicular, i.e., they form a right angle.
- Two vectors, x and y, in an inner product space, V, are orthogonal if their inner product $\langle x, y \rangle$ is zero. This relationship is denoted $x \perp y$
- Two vector subspaces, A and B, of an inner product space, V, are called **orthogonal subspaces** if each vector in A is orthogonal to each vector in B. The largest subspace of V that is orthogonal to a given subspace is its orthogonal complement.
- Given a module M and its dual M*, an element m′ of M* and an element m of M are orthogonal if their natural pairing is zero, i.e. ⟨m′, m⟩ = 0. Two sets S′ ⊆ M* and S ⊆ M are orthogonal if each element of S′ is orthogonal to each element of S.
- A term rewriting system is said to be orthogonal if it is left-linear and is non-ambiguous. Orthogonal term rewriting systems are confluent.

A set of vectors is called **pairwise orthogonal** if each pairing of them is orthogonal. Such a set is called an **orthogonal set**. Nonzero pairwise orthogonal vectors are always linearly independent.

In certain cases, the word normal is used to mean orthogonal, particularly in the geometric sense as in the normal to a surface. For example, the y-axis is normal to the curve $y = x^2$ at the origin. However, normal may also refer to the magnitude of a vector. In particular, a set is called orthonormal (orthogonal plus normal) if it is an orthogonal set of unit vectors. As a result, use of the term normal to mean "orthogonal" is often avoided. The word "normal" also has a different meaning in probability and statistics.

A vector space with a bilinear form generalizes the case of an inner product. When the bilinear form applied to two vectors results in zero, then they are **orthogonal**. The case of a pseudo-Euclidean plane uses the term hyperbolic orthogonality. In the diagram, axes x′ and t′ are hyperbolic-orthogonal for any given ϕ.

Euclidean Vector Spaces

In 2-D or higher-dimensional Euclidean space, two vectors are orthogonal if and only if their dot product is zero, i.e. they make an angle of 90°, or $\pi/2$ radians. Hence orthogonality of vectors is an extension of the concept of perpendicular vectors into higher-dimensional spaces.

In terms of Euclidean subspaces, a subspace has an "orthogonal complement" such that every vector in the subspace is orthogonal to every vector in the complement. In three-dimensional Euclidean space, the orthogonal complement of a line is the plane perpendicular to it, and vice versa.

Note however that there is no correspondence with regards to perpendicular planes, because vectors in subspaces start from the origin (by the definition of a Linear subspace).

In four-dimensional Euclidean space, the orthogonal complement of a line is a hyperplane and vice versa, and that of a plane is a plane.

Orthogonal Functions

By using integral calculus, it is common to use the following to define the inner product of two functions f and g:

$$\langle f, g\rangle_w = \int_a^b f(x)g(x)w(x)\,dx.$$

Here we introduce a nonnegative weight function w(x) in the definition of this inner product. In simple cases, w(x) = 1.

We say that these functions are **orthogonal** (with respect to this inner product) if the value of this integral is zero:

$$\int_a^b f(x)g(x)w(x)\,dx = 0.$$

Note that two functions which are orthogonal with respect to one inner product, aren't necessarily orthogonal with respect to another inner product.

We write the norms with respect to this inner product and the weight function as

$$\|f\|_w = \sqrt{\langle f, f\rangle_w}$$

The members of a set of functions { f_i : i = 1, 2, 3, ... } are:

- orthogonal on the closed interval [a, b] if

$$\langle f_i, f_j\rangle = \int_a^b f_i(x)f_j(x)w(x)\,dx = \|f_i\|^2\delta_{i,j} = \|f_j\|^2\delta_{i,j}$$

- orthonormal on the interval [a, b] if

$$\langle f_i, f_j\rangle = \int_a^b f_i(x)f_j(x)w(x)\,dx = \delta_{i,j}$$

where

$$\delta_{i,j} = \begin{cases} 1 & \text{if } i = j \\ 0 & \text{if } i \neq j \end{cases}$$

is the "Kronecker delta" function. In other words, any two of them are orthogonal, and the norm of each is 1 in the case of the orthonormal sequence. See in particular the orthogonal polynomials.

Examples

- The vectors $(1, 3, 2)^T$, $(3, -1, 0)^T$, $(1, 3, -5)^T$ are orthogonal to each other, since (1)(3) + (3)(−1) + (2)(0) = 0, (3)(1) + (−1)(3) + (0)(−5) = 0, and (1)(1) + (3)(3) + (2)(−5) = 0.
- The vectors $(1, 0, 1, 0, ...)^T$ and $(0, 1, 0, 1, ...)^T$ are orthogonal to each other. The dot product of these vectors is 0. We can then make the generalization to consider the vectors in $\mathbf{Z}_2^n$:

$$\mathbf{v}_k = \sum_{\substack{i=0 \\ ai+k<n}}^{n/a} \mathbf{e}_i$$

for some positive integer a, and for $1 \leq k \leq a - 1$, these vectors are orthogonal, for example $(1, 0, 0, 1, 0, 0, 1, 0)^T$, $(0, 1, 0, 0, 1, 0, 0, 1)^T$, $(0, 0, 1, 0, 0, 1, 0, 0)^T$ are orthogonal.

- The functions 2t + 3 and $45t^2 + 9t - 17$ are orthogonal with respect to a unit weight function on the interval from −1 to 1:

$$\int_{-1}^{1} (2t+3)\left(45t^2+9t-17\right)\,dt = 0$$

- The functions 1, sin(nx), cos(nx) : n = 1, 2, 3, ... are orthogonal with respect to Riemann integration on the intervals [0, 2π], [−π, π], or any other closed interval of length 2π. This fact is a central one in Fourier series.

Orthogonal Polynomials

- Various polynomial sequences named for mathematicians of the past are sequences of orthogonal polynomials. In particular:
  - The Hermite polynomials are orthogonal with respect to the Gaussian distribution with zero mean value.
  - The Legendre polynomials are orthogonal with respect to the uniform distribution on the interval [−1, 1].
  - The Laguerre polynomials are orthogonal with respect to the exponential distribution. Somewhat more general Laguerre polynomial sequences are orthogonal with respect to gamma distributions.
  - The Chebyshev polynomials of the first kind are orthogonal with respect to the measure $1/\sqrt{1-x^2}$. he Chebyshev polynomials of the second kind are orthogonal with respect to the Wigner semicircle distribution.

Orthogonal States in Quantum Mechanics

In quantum mechanics, a sufficient (but not necessary) condition that two eigenstates of a Hermitian operator ψ_m and ψ_n, are orthogonal is that they correspond to different eigenvalues. This means, in Dirac notation, that $\langle \psi_m | \psi_n \rangle = 0$ if ψ_m and ψ_n correspond to the same eigenvalue. This follows from the fact that Schrödinger's equation is a Sturm–Liouville equation (in Schrödinger's formulation) or that observables are given by hermitian operators (in Heisenberg's formulation).

Art

In art, the perspective (imaginary) lines pointing to the vanishing point are referred to as "orthogonal lines".

The term "orthogonal line" often has a quite different meaning in the literature of modern art criticism. Many works by painters such as Piet Mondrian and Burgoyne Diller are noted for their exclusive use of "orthogonal lines" — not, however, with reference to perspective, but rather referring to lines that are straight and exclusively horizontal or vertical, forming right angles where they intersect. For example, an essay at the Web site of the Thyssen-Bornemisza Museum states that "Mondrian ... dedicated his entire oeuvre to the investigation of the balance between orthogonal lines and primary colours."

Computer Science

Orthogonality in programming language design is the ability to use various language features in arbitrary combinations with consistent results.This usage was introduced by van Wijngaarden in the design of Algol 68:

The number of independent primitive concepts has been minimized in order that the language be easy to describe, to learn, and to implement. On the other hand, these concepts have been applied "orthogonally" in order to maximize the expressive power of the language while trying to avoid deleterious superfluities.

Orthogonality is a system design property which guarantees that modifying the technical effect produced by a component of a system neither creates nor propagates side effects to other components of the system. Typically this is achieved through the separation of concerns and encapsulation, and it is essential for feasible and compact designs of complex systems. The emergent behavior of a system consisting of components should be controlled strictly by formal definitions of its logic and not by side effects resulting from poor integration, i.e., non-orthogonal design of modules and interfaces. Orthogonality reduces testing and development time because it is easier to verify designs that neither cause side effects nor depend on them.

An instruction set is said to be **orthogonal** if it lacks redundancy (i.e., there is only a single instruction that can be used to accomplish a given task) and is designed such that instructions can use any register in any addressing mode. This terminology results from considering an instruction as a vector whose components are the instruction fields. One field identifies the registers to be operated upon and another specifies the addressing mode. An orthogonal instruction set uniquely encodes all combinations of registers and addressing modes.

Communications

In communications, multiple-access schemes are orthogonal when an ideal receiver can completely reject arbitrarily strong unwanted signals from the desired signal using different basis functions. One such scheme is TDMA, where the orthogonal basis functions are nonoverlapping rectangular pulses ("time slots").

Another scheme is orthogonal frequency-division multiplexing (OFDM), which refers to the use, by a single transmitter, of a set of frequency multiplexed signals with the exact minimum frequency spacing needed to make them orthogonal so that they do not interfere with each other. Well known examples include (**a**, **g**, and **n**) versions of 802.11 Wi-Fi; WiMAX; ITU-T G.hn, DVB-T, the terrestrial digital TV broadcast system used in most of the world outside North America; and DMT (Discrete Multi Tone), the standard form of ADSL.

In OFDM, the subcarrier frequencies are chosen so that the subcarriers are orthogonal to each other, meaning that crosstalk between the subchannels is eliminated and intercarrier guard bands are not required. This greatly simplifies the design of both the transmitter and the receiver. In conventional FDM, a separate filter for each subchannel is required.

Statistics, Econometrics, and Economics

When performing statistical analysis, independent variables that affect a particular dependent variable are said to be orthogonal if they are uncorrelated, since the covariance forms an inner product. In this case the same results are obtained for the effect of any of the independent variables upon the dependent variable, regardless of whether one models the effects of the variables individually with simple regression or simultaneously with multiple regression. If correlation is present, the factors are not orthogonal and different results are obtained by the two methods. This usage arises from the fact that if centered by subtracting the expected value (the mean), uncorrelated variables are orthogonal in the geometric sense discussed above, both as observed data (i.e., vectors) and as random variables (i.e., density functions). One econometric formalism that is alternative to the maximum likelihood framework, the Generalized Method of Moments, relies on orthogonality conditions. In particular, the Ordinary Least Squares estimator may be easily derived from an orthogonality condition between the explanatory variables and model residuals.

Taxonomy

In taxonomy, an orthogonal classification is one in which no item is a member of more than one group, that is, the classifications are mutually exclusive.

Combinatorics

In combinatorics, two n×n Latin squares are said to be orthogonal if their superimposition yields all possible n^2 combinations of entries.

Chemistry and Biochemistry

In synthetic organic chemistry orthogonal protection is a strategy allowing the deprotection of functional groups independently of each other. In chemistry and biochemistry, an orthogonal interaction occurs when there are two pairs of substances and each substance can interact with their respective partner, but does not interact with either substance of the other pair. For example, DNA has two orthogonal pairs: cytosine and guanine form a base-pair, and adenine and thymine form another base-pair, but other base-pair combinations are strongly disfavored. As a chemical example, tetrazine reacts with transcyclooctene and azide reacts with cyclooctyne without any cross-reaction, so these are mutually orthogonal reactions, and so, can be performed simultaneously and selectively. Bioorthogonal chemistry refers to chemical reactions occurring inside living systems without reacting with naturally present cellular components. In supramolecular chemistry the notion of orthogonality refers to the possibility of two or more supramolecular, often non-covalent, interactions being compatible; reversibly forming without interference from the other.

In analytical chemistry, analyses are "orthogonal" if they make a measurement or identification in completely different ways, thus increasing the reliability of the measurement. This is often required as a part of a new drug application.

System Reliability

In the field of system reliability orthogonal redundancy is that form of redundancy where the form of backup device or method is completely different from the prone to error device or method. The failure mode of an orthogonally redundant back-up device or method does not intersect with and is completely different from the failure mode of the device or method in need of redundancy to safeguard the total system against catastrophic failure.

Neuroscience

In neuroscience, a sensory map in the brain which has overlapping stimulus coding (e.g. location and quality) is called an orthogonal map.

Gaming

In board games such as chess which feature a grid of squares, 'orthogonal' is commonly used to mean "in the same row/'rank' or column/'file'". In this context 'orthogonal' and 'diagonal' are considered opposites. In the ancient Chinese board game Go a player can capture the stones of an opponent by occupying all orthogonally-adjacent points.

Other Examples

Stereo vinyl records encode both the left and right stereo channels in a single groove. By making the groove a 90-degree cut into the vinyl, variation in one wall was independent of variations in the other wall. The cartridge senses the resultant motion of the stylus following the groove in two orthogonal directions: 45 degrees from vertical to either side.

PARALLEL (GEOMETRY)

In geometry, **parallel** lines are lines in a plane which do not meet; that is, two lines in a plane that do not intersect or touch each other at any point are said to be parallel. By extension, a line and a plane, or two planes, in three-dimensional Euclidean space that do not share a point are said to be parallel. However, two lines in three-dimensional space which do not meet must be in a common plane to be considered parallel; otherwise they are called skew lines. Parallel planes are planes in the same three-dimensional space that never meet.

Parallel lines are the subject of Euclid's parallel postulate. Parallelism is primarily a property of affine geometries and Euclidean space is a special instance of this type of geometry. Some other spaces, such as hyperbolic space, have analogous properties that are sometimes referred to as parallelism.

Line art drawing of parallel lines and curves.

Symbol

The parallel symbol is ‖. For example, $AB \parallel CD$ indicates that line AB is parallel to line CD.

In the Unicode character set, the "parallel" and "not parallel" signs have codepoints U+2225 (∥) and U+2226 (∦), respectively. In addition, U+22D5 (⋕) represents the relation "equal and parallel to".

EUCLIDEAN PARALLELISM

Two Lines in a Plane

Conditions for Parallelism

Given parallel straight lines l and m in Euclidean space, the following properties are equivalent:

- Every point on line m is located at exactly the same (minimum) distance from line l (equidistant lines).
- Line m is in the same plane as line l but does not intersect l (recall that lines extend to infinity in either direction).
- When lines m and l are both intersected by a third straight line (a transversal) in the same plane, the corresponding angles of intersection with the transversal are congruent.

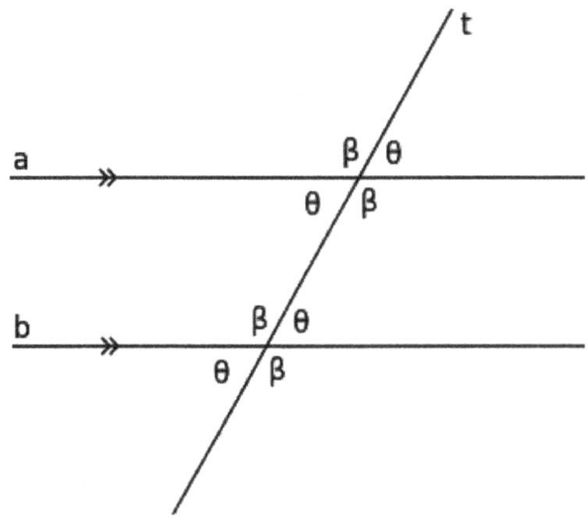

As shown by the tick marks, lines a and b are parallel. This can be proved because the transversal t produces congruent corresponding angles {\displaystyle \theta }, shown here both to the right of the transversal, one above and adjacent to line a and the other above and adjacent to line b.

Since these are equivalent properties, any one of them could be taken as the definition of parallel lines in Euclidean space, but the first and third properties involve measurement, and so, are more complicated than the second. Thus, the second property is the one usually chosen as the defining property of parallel lines in Euclidean geometry. The other properties are then consequences of Euclid's Parallel Postulate. Another property that also involves measurement is that lines parallel to each other have the same gradient (slope).

The definition of parallel lines as a pair of straight lines in a plane which do not meet appears as Definition 23 in Book I of Euclid's Elements. Alternative definitions were discussed by other Greeks, often as part of an attempt to prove the parallel postulate. Proclus attributes a definition of parallel lines as equidistant lines to Posidonius and quotes Geminus in a similar vein. Simplicius also mentions Posidonius' definition as well as its modification by the philosopher Aganis.

At the end of the nineteenth century, in England, Euclid's Elements was still the standard textbook in secondary schools. The traditional treatment of geometry was being pressured to change by the new developments in projective geometry and non-Euclidean geometry, so several new textbooks for the teaching of geometry were written at this time. A major difference between these reform texts, both between themselves and between them and Euclid, is the treatment of parallel lines. These reform texts were not without their critics and one of them, Charles Dodgson (a.k.a. Lewis Carroll), wrote a play, Euclid and His Modern Rivals, in which these texts are lambasted.

One of the early reform textbooks was James Maurice Wilson's Elementary Geometry of 1868. Wilson based his definition of parallel lines on the primitive notion of direction. According to Wilhelm Killing the idea may be traced back to Leibniz. Wilson, without defining direction since it is a primitive, uses the term in other definitions such as his sixth definition, "Two straight lines that meet one another have different directions, and the difference of their directions is the angle between them." Wilson (1868, p. 2) In definition 15 he introduces parallel lines in this way; "Straight lines which have the same direction, but are not parts of the same straight line, are called parallel lines." Wilson (1868, p. 12) Augustus De Morgan reviewed this text and declared it a failure, primarily on the basis of this definition and the way Wilson used it to prove things about parallel lines. Dodgson also devotes a large section of his play (Act II, Scene VI § 1) to denouncing Wilson's treatment of parallels. Wilson edited this concept out of the third and higher editions of his text.

Other properties, proposed by other reformers, used as replacements for the definition of parallel lines, did not fare much better. The main difficulty, as pointed out by Dodgson, was that to use them in this way required additional axioms to be added to the system. The equidistant line definition of Posidonius, expounded by Francis Cuthbertson in his 1874 text Euclidean Geometry suffers from the problem that the points that are found at a fixed given distance on one side of a straight line must be shown to form a straight line. This can not be proved and must be assumed to be true. The corresponding angles formed by a transversal property, used by W. D. Cooley in his 1860 text, The Elements of Geometry, simplified and explained requires a proof of the fact that if one transversal meets a pair of lines in congruent corresponding angles then all transversals must do so. Again, a new axiom is needed to justify this statement.

Construction

The three properties above lead to three different methods of construction of parallel lines.

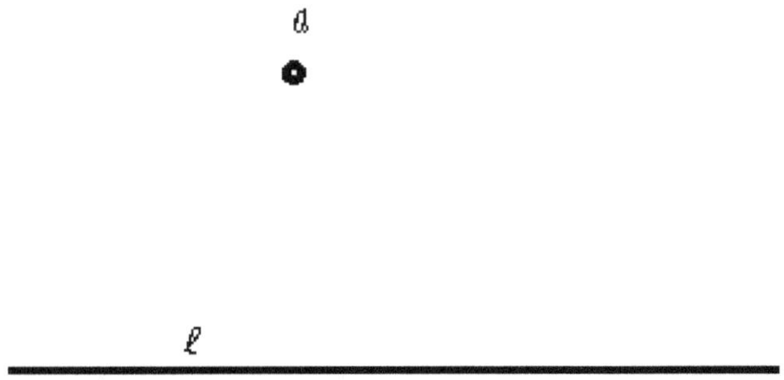

The problem: Draw a line through a parallel to l.

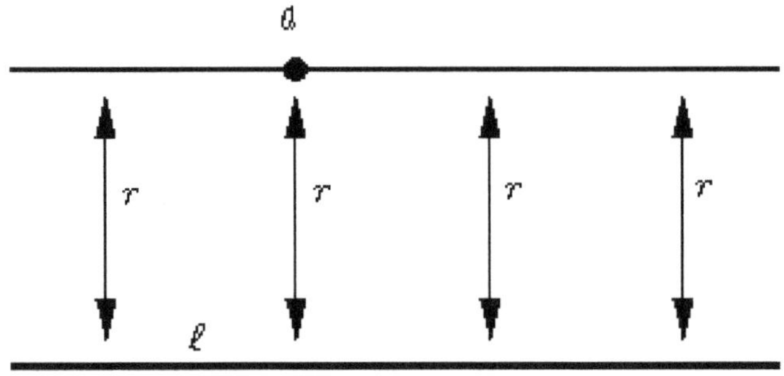

Property 1: Line m has everywhere the same distance to line l.

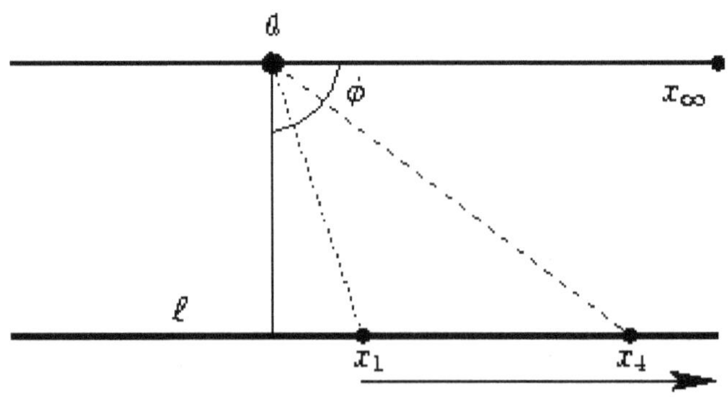

Property 2: Take a random line through a that intersects l in x. Move point x to infinity.

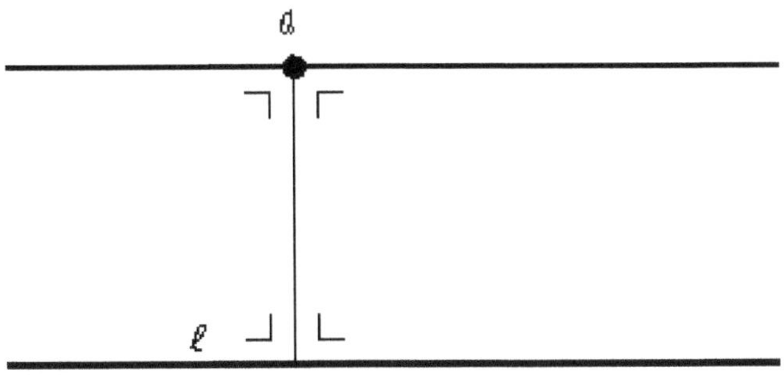

Property 3: Both l and m share a transversal line through a that intersect them at 90°.

Distance between two Parallel Lines

Because parallel lines in a Euclidean plane are equidistant there is a unique distance between the two parallel lines. Given the equations of two non-vertical, non-horizontal parallel lines,

$$y = mx + b_1$$
$$y = mx + b_2\ ,$$

the distance between the two lines can be found by locating two points (one on each line) that lie on a common perpendicular to the parallel lines and calculating the distance between them. Since the lines have slope m, a common perpendicular would have slope −1/m and we can take the line with equation y = −x/m as a common perpendicular. Solve the linear systems

$$\begin{cases} y = mx + b_1 \\ y = -x/m \end{cases}$$

and

$$\begin{cases} y = mx + b_2 \\ y = -x/m \end{cases}$$

to get the coordinates of the points. The solutions to the linear systems are the points

$$(x_1, y_1) = \left(\frac{-b_1 m}{m^2+1}, \frac{b_1}{m^2+1}\right)$$

and

$$(x_2, y_2) = \left(\frac{-b_2 m}{m^2+1}, \frac{b_2}{m^2+1}\right).$$

These formulas still give the correct point coordinates even if the parallel lines are horizontal (i.e., m = 0). The distance between the points is

$$d = \sqrt{\left(\frac{b_1 m - b_2 m}{m^2 + 1}\right)^2 + \left(\frac{b_2 - b_1}{m^2 + 1}\right)^2},$$

which reduces to

$$d = \frac{|b_2 - b_1|}{\sqrt{m^2 + 1}}.$$

When the lines are given by the general form of the equation of a line (horizontal and vertical lines are included):

$$ax + by + c_1 = 0$$
$$ax + by + c_2 = 0,$$

their distance can be expressed as

$$d = \frac{|c_2 - c_1|}{\sqrt{a^2 + b^2}}.$$

Two Lines in Three-Dimensional Space

Two lines in the same three-dimensional space that do not intersect need not be parallel. Only if they are in a common plane are they called parallel; otherwise they are called skew lines.

Two distinct lines l and m in three-dimensional space are parallel if and only if the distance from a point P on line m to the nearest point on line l is independent of the location of P on line m. This never holds for skew lines.

A Line and a Plane

A line m and a plane q in three-dimensional space, the line not lying in that plane, are parallel if and only if they do not intersect.

Equivalently, they are parallel if and only if the distance from a point P on line m to the nearest point in plane q is independent of the location of P on line m.

Two Planes

Similar to the fact that parallel lines must be located in the same plane, parallel planes must be situated in the same three-dimensional space and contain no point in common.

Two distinct planes q and r are parallel if and only if the distance from a point P in plane q to the nearest point in plane r is independent of the location of P in plane q. This will never hold if the two planes are not in the same three-dimensional space.

Extension to Non-Euclidean Geometry

In non-Euclidean geometry, it is more common to talk about geodesics than (straight) lines. A geodesic is the shortest path between two points in a given geometry. In physics this may be interpreted as the path that a particle follows if no force is applied to it. In non-Euclidean geometry (elliptic or hyperbolic geometry) the three Euclidean properties mentioned above are not equivalent and only the second one,(Line m is in the same plane as line l but does not intersect l) since it involves no measurements is useful in non-Euclidean geometries. In general geometry the three properties above give three different types of curves, **equidistant curves**, **parallel geodesics** and **geodesics sharing a common perpendicular**, respectively.

Hyperbolic Geometry

While in Euclidean geometry two geodesics can either intersect or be parallel, in hyperbolic geometry, there are three possibilities. Two geodesics belonging to the same plane can either be:

- **intersecting**, if they intersect in a common point in the plane,
- **parallel**, if they do not intersect in the plane, but converge to a common limit point at infinity (ideal point), or
- **ultra parallel**, if they do not have a common limit point at infinity.

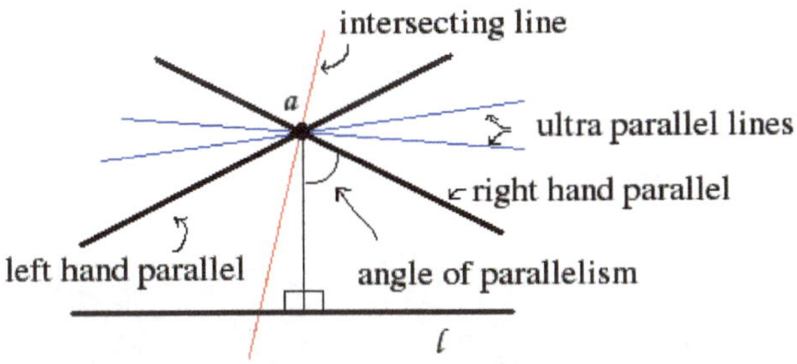

Intersecting, **parallel** and **ultra parallel** lines through a with respect to l in the hyperbolic plane. The parallel lines appear to intersect l just off the image. This is just an artifact of the visualisation. On a real hyperbolic plane the lines will get closer to each other and ‹meet' in infinity.

In the literature ultra parallel geodesics are often called non-intersecting. Geodesics intersecting at infinity are called limiting parallel.

As in the illustration through a point a not on line l there are two limiting parallel lines, one for each direction ideal point of line l. They separate the lines intersecting line l and those that are ultra parallel to line l.

Ultra parallel lines have single common perpendicular (ultraparallel theorem), and diverge on both sides of this common perpendicular.

Spherical or Elliptic Geometry

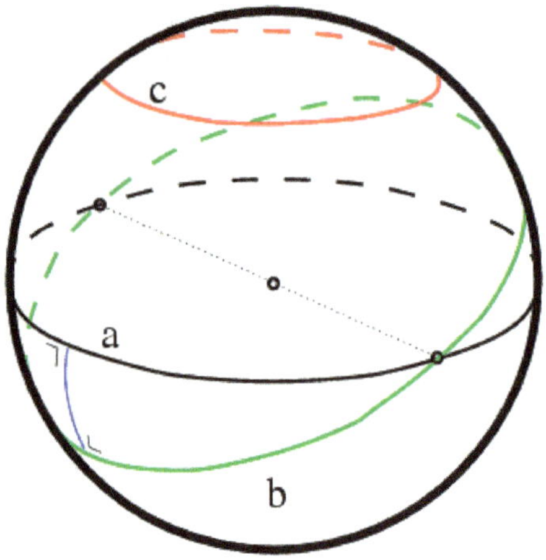

On the sphere there is no such thing as a parallel line. Line a is a great circle, the equivalent of a straight line in spherical geometry. Line c is equidistant to line a but is not a great circle. It is a parallel of latitude. Line b is another geodesic which intersects a in two antipodal points. They share two common perpendiculars (one shown in blue).

In spherical geometry, all geodesics are great circles. Great circles divide the sphere in two equal hemispheres and all great circles intersect each other. Thus, there are no parallel geodesics to a given geodesic, as all geodesics intersect. Equidistant curves on the sphere are called **parallels of latitude** analogous to the latitude lines on a globe. Parallels of latitude can be generated by the intersection of the sphere with a plane parallel to a plane through the center of the sphere.

Reflexive Variant

In synthetic, affine geometry the relation of two parallel lines is a fundamental concept that is modified from the usage in Euclidean geometry. It is clear that the relation of parallelism is a symmetric relation and a transitive relation. These are two properties of an equivalence relation. In Euclidean geometry a line is not considered to be parallel to itself, but in affine geometry it is convenient to hold a line as parallel to itself, thus yielding parallelism as an equivalence relation.

Another way of describing this type of parallelism is the requirement that their intersection is not a singleton. Two lines are then parallel when they have all or none of their points in common. It has been noted that Playfair's axiom used in affine and Euclidean geometry is then equivalent to the statement that parallelism forms a transitive relation on the set of lines in the plane.

Vertex (Geometry)

In geometry, a **vertex** (plural: **vertices** or **vertexes**) is a point where two or more curves, lines, or edges meet. As a consequence of this definition, the point where two lines meet to form an angle and the corners of polygons and polyhedra are vertices.

Definition

Of an Angle

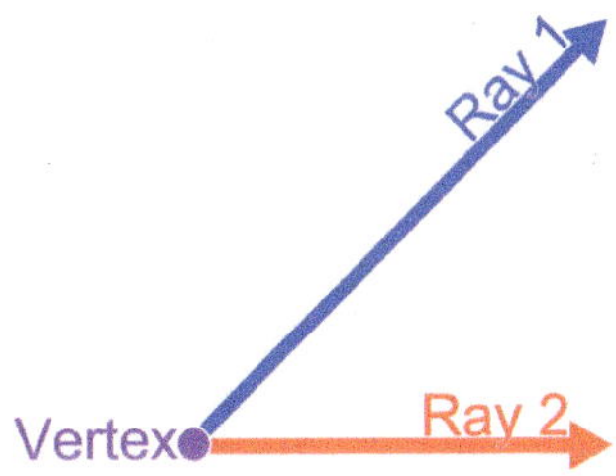

A vertex of an angle is the endpoint where two line segments or rays come together.

The vertex of an angle is the point where two rays begin or meet, where two line segments join or meet, where two lines intersect (cross), or any appropriate combination of rays, segments and lines that result in two straight "sides" meeting at one place.

Of a Polytope

A vertex is a corner point of a polygon, polyhedron, or other higher-dimensional polytope, formed by the intersection of edges, faces or facets of the object.

In a polygon, a vertex is called "convex" if the internal angle of the polygon, that is, the angle formed by the two edges at the vertex, with the polygon inside the angle, is less than π radians (180°, two right angles) ; otherwise, it is called "concave" or "reflex". More generally, a vertex of a polyhedron or polytope is convex if the intersection of the polyhedron or polytope with a sufficiently small sphere centered at the vertex is convex, and concave otherwise. Polytope vertices are related to vertices of graphs, in that the 1-skeleton of a polytope is a graph, the vertices of which correspond to the vertices of the polytope, and in that a graph can be viewed as a 1-dimensional simplicial complex the vertices of which are the graph's vertices. However, in graph theory, vertices may have fewer than two incident edges, which is usually not allowed for geometric vertices. There is also a connection between geometric vertices and the vertices of a curve, its points of extreme curvature: in some sense the vertices of a polygon are points of infinite curvature, and if a polygon is approximated by a smooth curve there will be a point of extreme curvature near each polygon vertex.However, a smooth curve approximation to a polygon will also have additional vertices, at the points where its curvature is minimal.

Of a Plane Tiling

A vertex of a plane tiling or tessellation is a point where three or more tiles meet; generally, but not always, the tiles of a tessellation are polygons and the vertices of the tessellation are also vertices of its tiles. More generally, a tessellation can be viewed as a kind of topological cell complex, as can the faces of a polyhedron or polytope; the vertices of other kinds of complexes such as simplicial complexes are its zero-dimensional faces.

Principal Vertex

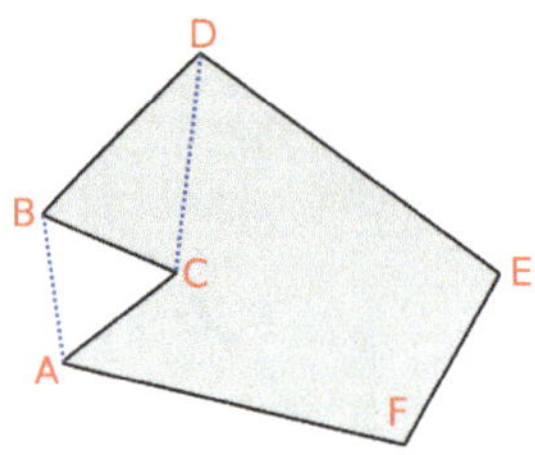

Vertex B is an ear, because the open line segment between C and D is entirely inside the polygon. Vertex C is a mouth, because the open line segment between A and B is entirely outside the polygon.

A polygon vertex x_i of a simple polygon P is a principal polygon vertex if the diagonal $[x_{(i-1)}, x_{(i+1)}]$intersects the boundary of P only at $x_{(i-1)}$ and $x_{(i+1)}$. There are two types of principal vertices: ears and mouths.

Ears

A principal vertex x_i of a simple polygon P is called an ear if the diagonal $[x_{(i-1)}, x_{(i+1)}]$ that bridges x_i lies entirely in P. According to the two ears theorem, every simple polygon has at least two ears.

Mouths

A principal vertex x_i of a simple polygon P is called a mouth if the diagonal $[x_{(i-1)}, x_{(i+1)}]$ lies outside the boundary of P.

Number of Vertices of a Polyhedron

Any convex polyhedron's surface has Euler characteristic

$$V - E + F = 2,$$

where V is the number of vertices, E is the number of edges, and F is the number of faces. This equation is known as Euler's polyhedron formula. Thus the number of vertices is 2 more than the excess of the number of edges over the number of faces. For example, a cube has 12 edges and 6 faces, and hence 8 vertices.

Vertices in Computer Graphics

In computer graphics, objects are often represented as triangulated polyhedra in which the object vertices are associated not only with three spatial coordinates but also with other graphical information necessary to render the object correctly, such as colors, reflectance properties, textures, and surface normals; these properties are used in rendering by a vertex shader, part of the vertex pipeline.

Congruence (Geometry)

In geometry, two figures or objects are **congruent** if they have the same shape and size, or if one has the same shape and size as the mirror image of the other.

More formally, two sets of points are called **congruent** if, and only if, one can be transformed into the other by an isometry, i.e., a combination of **rigid motions**, namely a translation, a rotation, and a reflection. This means that either object can be repositioned and reflected (but not resized) so as to coincide precisely with the other object. So two distinct plane figures on a piece of paper are congruent if we can cut them out and then match them up completely. Turning the paper over is permitted.

In elementary geometry the word congruent is often used as follows. The word equal is often used in place of congruent for these objects.

- Two line segments are congruent if they have the same length.
- Two angles are congruent if they have the same measure.
- Two circles are congruent if they have the same diameter.

In this sense, two plane figures are congruent implies that their corresponding characteristics are "congruent" or "equal" including not just their corresponding sides and angles, but also their corresponding diagonals, perimeters and areas.

The related concept of similarity applies if the objects differ in size but not in shape.

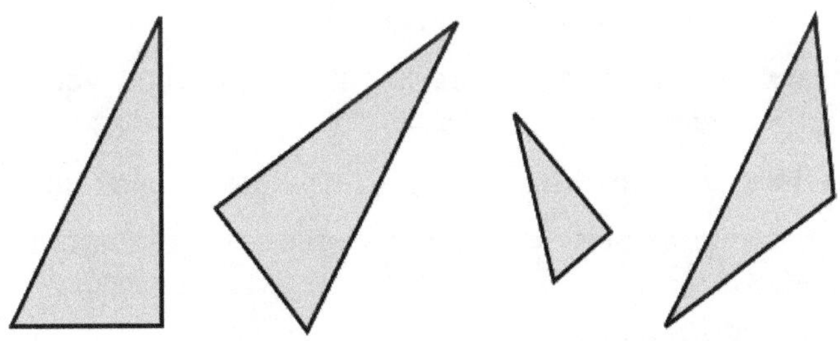

An example of congruence. The two triangles on the left are congruent, while the third is similar to them. The last triangle is neither similar nor congruent to any of the others. Note that congruence permits alteration of some properties, such as location and orientation, but leaves others unchanged, like distance and angles. The unchanged properties are called invariants.

Determining Congruence of Polygons

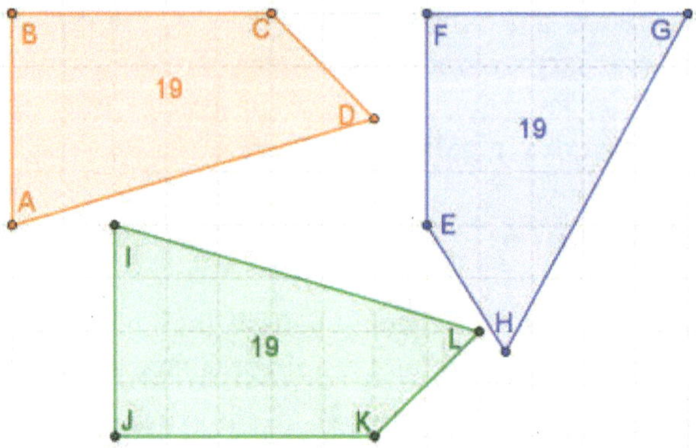

The orange and green quadrilaterals are congruent; the blue is not congruent to them. All three have the same perimeter and area. (The ordering of the sides of the blue quadrilateral is "mixed" which results in two of the interior angles and one of the diagonals not being congruent.)

For two polygons to be congruent, they must have an equal number of sides (and hence an equal number—the same number—of vertices). Two polygons with n sides are congruent if and only if they each have numerically identical sequences (even if clockwise for one polygon and counterclockwise for the other) side-angle-side-angle-... for n sides and n angles.

Congruence of polygons can be established graphically as follows:

- First, match and label the corresponding vertices of the two figures.
- Second, draw a vector from one of the vertices of the one of the figures to the corresponding vertex of the other figure. Translate the first figure by this vector so that these two vertices match.
- Third, rotate the translated figure about the matched vertex until one pair of corresponding sides matches.
- Fourth, reflect the rotated figure about this matched side until the figures match.

If at any time the step cannot be completed, the polygons are not congruent.

Congruence of Triangles

Two triangles are congruent if their corresponding sides are equal in length, in which case their corresponding angles are equal in measure.

If triangle ABC is congruent to triangle DEF, the relationship can be written mathematically as:

$$\triangle ABC \cong \triangle DEF$$

In many cases it is sufficient to establish the equality of three corresponding parts and use one of the following results to deduce the congruence of the two triangles.

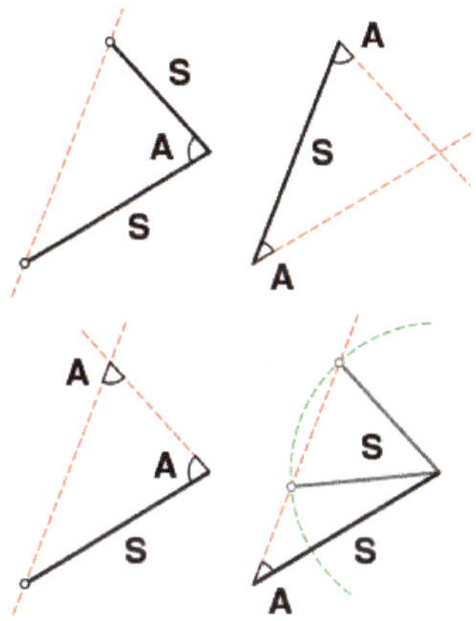

The shape of a triangle is determined up to congruence by specifying two sides and the angle between them (SAS), two angles and the side between them (ASA) or two angles and a corresponding adjacent side (AAS). Specifying two sides and an adjacent angle (SSA), however, can yield two distinct possible triangles.

Determining Congruence

Sufficient evidence for congruence between two triangles in Euclidean space can be shown through the following comparisons:

- **SAS** (Side-Angle-Side): If two pairs of sides of two triangles are equal in length, and the included angles are equal in measurement, then the triangles are congruent.
- **SSS** (Side-Side-Side): If three pairs of sides of two triangles are equal in length, then the triangles are congruent.
- **ASA** (Angle-Side-Angle): If two pairs of angles of two triangles are equal in measurement, and the included sides are equal in length, then the triangles are congruent. The ASA Postulate was contributed by Thales of Miletus (Greek). In most systems of axioms, the three criteria—**SAS**, **SSS** and **ASA**—are established as theorems. In the School Mathematics Study Group system **SAS** is taken as one (#15) of 22 postulates.
- **AAS** (Angle-Angle-Side): If two pairs of angles of two triangles are equal in measurement, and a pair of ***corresponding*** non-included sides are equal in length, then the triangles are congruent. (In British usage, ***ASA*** and ***AAS*** are usually combined into a single condition ***AAcorrS*** - any two angles and a corresponding side.) For American usage, AAS is equivalent to an ASA condition, by the fact that if any two angles are given, so is the third angle, since their sum should

be 180°.

- **RHS** (Right-angle-Hypotenuse-Side), also known as **HL** (Hypotenuse-Leg): If two right-angled triangles have their hypotenuses equal in length, and a pair of shorter sides are equal in length, then the triangles are congruent.

Side-Side-Angle

The SSA condition (Side-Side-Angle) which specifies two sides and a non-included angle (also known as ASS, or Angle-Side-Side) does not by itself prove congruence. In order to show congruence, additional information is required such as the measure of the corresponding angles and in some cases the lengths of the two pairs of corresponding sides. There are a few possible cases:

If two triangles satisfy the SSA condition and the length of the side opposite the angle is greater than or equal to the length of the adjacent side (SsA, or long side-short side-angle), then the two triangles are congruent. The opposite side is sometimes longer when the corresponding angles are acute, but it is always longer when the corresponding angles are right or obtuse. Where the angle is a right angle, also known as the Hypotenuse-Leg (HL) postulate or the Right-angle-Hypotenuse-Side (RHS) condition, the third side can be calculated using the Pythagorean Theorem thus allowing the SSS postulate to be applied.

If two triangles satisfy the SSA condition and the corresponding angles are acute and the length of the side opposite the angle is equal to the length of the adjacent side multiplied by the sine of the angle, then the two triangles are congruent.

If two triangles satisfy the SSA condition and the corresponding angles are acute and the length of the side opposite the angle is greater than the length of the adjacent side multiplied by the sine of the angle (but less than the length of the adjacent side), then the two triangles cannot be shown to be congruent. This is the ambiguous case and two different triangles can be formed from the given information.

Angle-Angle-Angle

In Euclidean geometry, **AAA** (Angle-Angle-Angle) (or just **AA**, since in Euclidean geometry the angles of a triangle add up to 180°) does not provide information regarding the size of the two triangles and hence proves only similarity and not congruence in Euclidean space.

However, in spherical geometry and hyperbolic geometry (where the sum of the angles of a triangle varies with size) **AAA** is sufficient for congruence on a given curvature of surface.

Definition of Congruence in Analytic Geometry

In a Euclidean system, congruence is fundamental; it is the counterpart of equality for numbers. In analytic geometry, congruence may be defined intuitively thus: two mappings of figures onto one Cartesian coordinate system are congruent if and only if, for any two points in the first mapping, the Euclidean distance between them is equal to the Euclidean distance between the corresponding points in the second mapping.

A more formal definition states that two subsets A and B of Euclidean space $\mathbf{R}^n$ are called congruent if there exists an isometry f : $\mathbf{R}^n \rightarrow \mathbf{R}^n$ (an element of the Euclidean group E(n)) with f(A) = B. Congruence is an equivalence relation.

Congruent Conic Sections

Two conic sections are congruent if their eccentricities and one other distinct parameter characterizing them are equal. Their eccentricities establish their shapes, equality of which is sufficient to establish similarity, and the second parameter then establishes size. Since two circles, parabolas, or rectangular hyperbolas always have the same eccentricity (specifically 0 in the case of circles, 1 in the case of parabolas, and $\sqrt{2}$ in the case of rectangular hyperbolas), two circles, parabolas, or rectangular hyperbolas need to have only one other common parameter value, establishing their size, for them to be congruent.

Congruent Polyhedra

For two polyhedra with the same number E of edges, the same number of faces, and the same number of sides on corresponding faces, there exists a set of at most E measurements that can establish whether or not the polyhedra are congruent. For cubes, which have 12 edges, only 9 measurements are necessary.

Congruent Triangles on a Sphere

As with plane triangles, on a sphere two triangles sharing the same sequence of angle-side-angle (ASA) are necessarily congruent (that is, they have three identical sides and three identical angles). This can be seen as follows: One can situate one of the vertices with a given angle at the south pole and run the side with given length up the prime meridian. Knowing both angles at either end of the segment of fixed length ensures that the other two sides emanate with a uniquely determined trajectory, and thus will meet each other at a uniquely determined point; thus ASA is valid.

The congruence theorems side-angle-side (SAS) and side-side-side (SSS) also hold on a sphere; in addition, if two spherical triangles have an identical angle-angle-angle (AAA) sequence, they are congruent (unlike for plane triangles).

The plane-triangle congruence theorem angle-angle-side (AAS) does not hold for spherical triangles. As in plane geometry, side-side-angle (SSA) does not imply congruence.

SIMILARITY (GEOMETRY)

Two geometrical objects are called **similar** if they both have the same shape, or one has the same shape as the mirror image of the other. More precisely, one can be obtained from the other by uniformly scaling (enlarging or reducing), possibly with additional translation, rotation and reflection. This means that either object can be rescaled, repositioned, and reflected, so as to coincide precisely with the other object. If two objects are similar, each is congruent to the result of a particular uniform scaling of the other. A modern and novel perspective of similarity is to consider geometrical objects similar if one appears congruent to the other when zoomed in or out at some level.

For example, all circles are similar to each other, all squares are similar to each other, and all equilateral triangles are similar to each other. On the other hand, ellipses are not all similar to each other, rectangles are not all similar to each other, and isosceles triangles are not all similar to each other.

If two angles of a triangle have measures equal to the measures of two angles of another triangle, then the triangles are similar. Corresponding sides of similar polygons are in proportion, and corresponding angles of similar polygons have the same measure.

This article assumes that a scaling can have a scale factor of 1, so that all congruent shapes are also similar, but some school textbooks specifically exclude congruent triangles from their definition of similar triangles by insisting that the sizes must be different if the triangles are to qualify as similar.

Figures shown in the same color are similar

Similar Triangles

In geometry two triangles, ΔABC and ΔA⊠B⊠C⊠, are similar if and only if corresponding angles have the same measure: this implies that they are similar if and only if the lengths of corresponding sides are proportional. It can be shown that two triangles having congruent angles (equiangular triangles) are similar, that is, the corresponding sides can be proved to be proportional. This is known as the AAA similarity theorem. Due to this theorem, several authors simplify the definition of similar triangles to only require that the corresponding three angles are congruent.

There are several statements each of which is necessary and sufficient for two triangles to be similar:

1. The triangles have two congruent angles, which in Euclidean geometry implies that all their angles are congruent. That is:

If $\angle BAC$ is equal in measure to $\angle B'A'C'$, and $\angle ABC$ is equal in measure to $\angle A'B'C'$, then this implies that $\angle ACB$ is equal in measure to $\angle A'C'B'$ and the triangles are similar.

2. All the corresponding sides have lengths in the same ratio:

   $AB/A'B' = BC/B'C' = AC/A'C'$. This is equivalent to saying that one triangle (or its mirror image) is an enlargement of the other.

3. Two sides have lengths in the same ratio, and the angles included between these sides have the same measure. For instance:

$AB/A'B' = BC/B'C'$ and $\angle ABC$ is equal in measure to $\angle A'B'C'$.

This is known as the SAS Similarity Criterion.

When two triangles $\triangle ABC$ and $\triangle A'B'C'$ are similar, one writes[p. 22]

$\triangle ABC \sim \triangle A'B'C'$.

There are several elementary results concerning similar triangles in Euclidean geometry:

- Any two equilateral triangles are similar.
- Two triangles, both similar to a third triangle, are similar to each other (transitivity of similarity of triangles).
- Corresponding altitudes of similar triangles have the same ratio as the corresponding sides.
- Two right triangles are similar if the hypotenuse and one other side have lengths in the same ratio.

Given a triangle $\triangle ABC$ and a line segment DE one can, with straightedge and compass, find a point F such that $\triangle ABC \sim \triangle DEF$. The statement that the point F satisfying this condition exists is Wallis's Postulate and is logically equivalent to Euclid's Parallel Postulate. In hyperbolic geometry (where Wallis's Postulate is false) similar triangles are congruent.

In the axiomatic treatment of Euclidean geometry given by G.D. Birkhoff (see Birkhoff's axioms) the SAS Similarity Criterion given above was used to replace both Euclid's Parallel Postulate and the SAS axiom which enabled the dramatic shortening of Hilbert's axioms.

Other Similar Polygons

The concept of similarity extends to polygons with more than three sides. Given any two similar polygons, corresponding sides taken in the same sequence (even if clockwise for one polygon and counterclockwise for the other) are proportional and corresponding angles taken in the same sequence are equal in measure. However, proportionality of corresponding sides is not by itself sufficient to prove similarity for polygons beyond triangles (otherwise, for example, all rhombi would be similar). Likewise, equality of all angles in sequence is not sufficient to guarantee similarity (otherwise all rectangles would be similar). A sufficient condition for similarity of polygons is that corresponding sides and diagonals are proportional.

For given n, all regular n-gons are similar.

Similar Curves

Several types of curves have the property that all examples of that type are similar to each other. These include:

- Circles
- Parabolas
- Hyperbolas of a specific eccentricity
- Ellipses of a specific eccentricity
- Catenaries
- Graphs of the logarithm function for different bases
- Graphs of the exponential function for different bases
- Logarithmic spirals

Similarity in Euclidean Space

A **similarity** (also called a **similarity transformation** or **similitude**) of a Euclidean space is a bijection f from the space onto itself that multiplies all distances by the same positive real number r, so that for any two points x and y we have

$$d(f(x), f(y)) = rd(x, y),$$

where "d(x,y)" is the Euclidean distance from x to y. The scalar r has many names in the literature including; the ratio of similarity, the stretching factor and the similarity coefficient. When r = 1 a similarity is called an isometry (rigid motion). Two sets are called **similar** if one is the image of the other under a similarity.

As a map $f : \mathbb{R}^n \to \mathbb{R}^n$, a similarity of ratio r takes the form

$$f(x) = rAx + t,$$

where $A \in O_n(\mathbb{R})$ is an n × n orthogonal matrix and $t \in \mathbb{R}^n$ is a translation vector.

Similarities preserve planes, lines, perpendicularity, parallelism, midpoints, inequalities between distances and line segments. Similarities preserve angles but do not necessarily preserve orientation, direct similitudes preserve orientation and opposite similitudes change it.

The similarities of Euclidean space form a group under the operation of composition called the similarities group S. The direct similitudes form a normal subgroup of S and the Euclidean group E(n) of isometries also forms a normal subgroup. The similarities group S is itself a subgroup of the affine group, so every similarity is an affine transformation.

One can view the Euclidean plane as the complex plane, that is, as a 2-dimensional space over the reals. The 2D similarity transformations can then be expressed in terms of complex arithmetic and are given by f(z) = az + b (direct similitudes) and f(z) = az + b (opposite similitudes), where a and b are complex numbers, $a \neq 0$. When $|a| = 1$, these similarities are isometries.

Ratios of Sides, of Areas, and of Volumes

The ratio between the areas of similar figures is equal to the square of the ratio of corresponding lengths of those figures (for example, when the side of a square or the radius of a circle is multiplied by three, its area is multiplied by nine — i.e. by three squared). The altitudes of similar triangles are in the same ratio as corresponding sides. If a triangle has a side of length b and an altitude drawn to that side of length h then a similar triangle with corresponding side of length kb will have an altitude drawn to that side of length kh. The area of the first triangle is, A = 1/2bh, while the area of the similar triangle will be $A' = 1/2(kb)(kh) = k^2A$. Similar figures which can be decomposed into similar triangles will have areas related in the same way. The relationship holds for figures that are not rectifiable as well.

The ratio between the volumes of similar figures is equal to the cube of the ratio of corresponding lengths of those figures (for example, when the edge of a cube or the radius of a sphere is multiplied by three, its volume is multiplied by 27 — i.e. by three cubed).

Galileo's square–cube law concerns similar solids. If the ratio of similitude (ratio of corresponding sides) between the solids is k, then the ratio of surface areas of the solids will be k^2, while the ratio of volumes will be k^3.

Similarity in General Metric Spaces

Sierpiński triangle. A space having self-similarity dimension log 3/log 2 = $\log_2 3$, which is approximately 1.58. (From Hausdorff dimension.)

In a general metric space (X, d), an exact **similitude** is a function f from the metric space Xinto itself that multiplies all distances by the same positive scalar r, called f 's contraction factor, so that for any two points x and y we have

$$d(f(x), f(y)) = rd(x, y).$$

Weaker versions of similarity would for instance have f be a bi-Lipschitz function and the scalar r a limit

$$\lim \frac{d(f(x), f(y))}{d(x, y)} = r.$$

This weaker version applies when the metric is an effective resistance on a topologically self-similar set.

A self-similar subset of a metric space (X, d) is a set K for which there exists a finite set of similitudes $\{f_s\}_{s \in S}$ with contraction factors $0 \leq r_s < 1$ such that K is the unique compact subset of X for which

$$\bigcup_{s \in S} f_s(K) = K.$$

These self-similar sets have a self-similar measure μ^D with dimension D given by the formula

$$\sum_{s \in S} (r_s)^D = 1$$

which is often (but not always) equal to the set's Hausdorff dimension and packing dimension. If the overlaps between the $f_s(K)$ are "small", we have the following simple formula for the measure:

$$\mu^D(f_{s_1} \circ f_{s_2} \circ \cdots \circ f_{s_n}(K)) = (r_{s_1} \cdot r_{s_2} \cdots r_{s_n})^D.$$

Topology

In topology, a metric space can be constructed by defining a **similarity** instead of a distance. The similarity is a function such that its value is greater when two points are closer (contrary to the distance, which is a measure of **dissimilarity:** the closer the points, the lesser the distance).

The definition of the similarity can vary among authors, depending on which properties are desired. The basic common properties are

- Positive defined:

$$\forall (a, b), S(a, b) \geq 0$$

Majored by the similarity of one element on itself (**auto-similarity**):

$$S(a, b) \leq S(a, a) \quad \text{and} \quad \forall (a, b), S(a, b) = S(a, a) \Leftrightarrow a = b$$

More properties can be invoked, such as reflectivity ($\forall (a, b)\ S(a, b) = S(b, a)$) or **finiteness** ($\forall (a, b)\ S(a, b) < \infty$).

The upper value is often set at 1 (creating a possibility for a probabilistic interpretation of the similitude).

Self-Similarity

Self-similarity means that a pattern is non-trivially similar to itself, e.g., the set {..., 0.5, 0.75, 1, 1.5, 2, 3, 4, 6, 8, 12, ...} of numbers of the form $\{2^i, 3\cdot2^i\}$ where i ranges over all integers. When this set is plotted on a logarithmic scale it has one-dimensional translational symmetry: adding or subtracting the logarithm of two to the logarithm of one of these numbers produces the logarithm of another of these numbers. In the given set of numbers themselves, this corresponds to a similarity transformation in which the numbers are multiplied or divided by two.

Psychology

The intuition for the notion of geometric similarity already appears in human children, as can be seen in their drawings.

SYMMETRY

Symmetry (from Greek συμμετρία symmetria "agreement in dimensions, due proportion, arrangement") in everyday language refers to a sense of harmonious and beautiful proportion and balance. In mathematics, "symmetry" has a more precise definition, that an object is invariant to any of various transformations; including reflection, rotation or scaling. Although these two meanings of "symmetry" can sometimes be told apart, they are related, so they are here discussed together.

Mathematical symmetry may be observed with respect to the passage of time; as a spatial relationship; through geometric transformations; through other kinds of functional transformations; and as an aspect of abstract objects, theoretic models, language, music and even knowledge itself.

This article describes symmetry from three perspectives: in mathematics, including geometry, the most familiar type of symmetry for many people; in science and nature; and in the arts, covering architecture, art and music.

The opposite of symmetry is asymmetry.

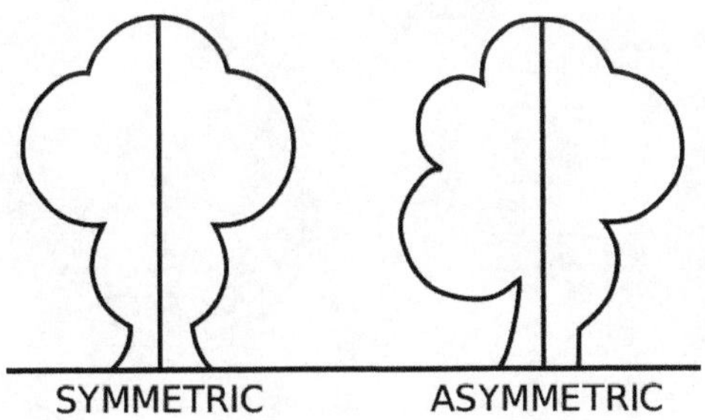

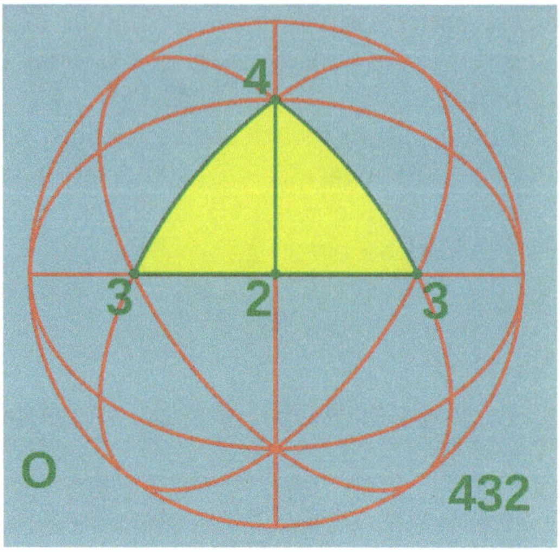

Sphere symmetrical group of representing an octahedral rotational symmetry. The yellow region shows the fundamental domain.

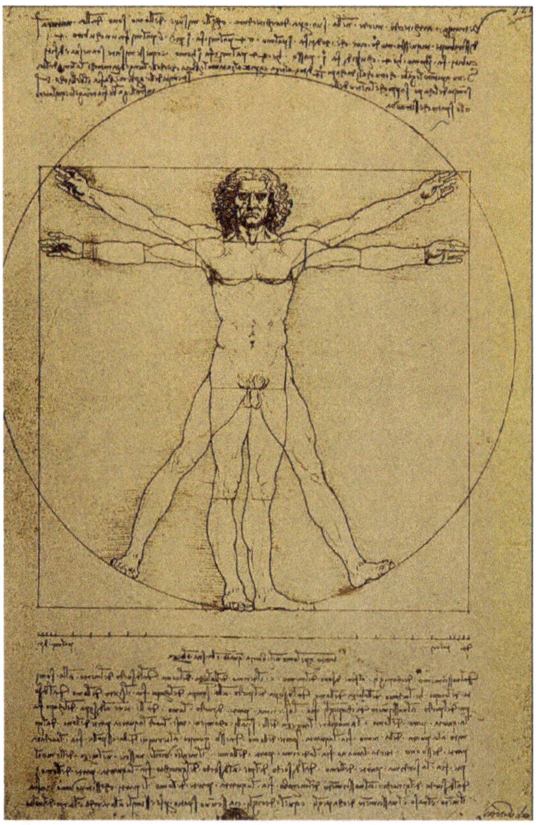

Leonardo da Vinci's 'Vitruvian Man' (ca. 1487) is often used as a representation of symmetry in the human body and, by extension, the natural universe.

A fractal-like shape that has reflectional symmetry, rotational symmetry and self-similarity, three forms of symmetry. This shape is obtained by a finite subdivision rule.

Symmetric arcades of a portico in the Great Mosque of Kairouan also called the Mosque of Uqba, in Tunisia.

In Mathematics

In Geometry

The triskelion has 3-fold rotational symmetry.

A geometric shape or object is symmetric if it can be divided into two or more identical pieces that are arranged in an organized fashion. This means that an object is symmetric if there is a transformation that moves individual pieces of the object but doesn't change the overall shape. The type of symmetry is determined by the way the pieces are organized, or by the type of transformation:

- An object has reflectional symmetry (line or mirror symmetry) if there is a line going through it which divides it into two pieces which are mirror images of each other.
- An object has rotational symmetry if the object can be rotated about a fixed point without changing the overall shape.
- An object has translational symmetry if it can be translated without changing its overall shape.
- An object has helical symmetry if it can be simultaneously translated and rotated in three-dimensional space along a line known as a screw axis.
- An object has scale symmetry if it does not change shape when it is expanded or contracted. Fractals also exhibit a form of scale symmetry, where small portions of the fractal are similar in shape to large portions.
- Other symmetries include glide reflection symmetry and rotoreflection symmetry.

In Logic

A dyadic relation R is symmetric if and only if, whenever it's true that Rab, it's true that Rba. Thus, "is the same age as" is symmetrical, for if Paul is the same age as Mary, then Mary is the same age as Paul.

Symmetric binary logical connectives are and (∧, or &), or (∨, or |), biconditional (if and only if) (↔), nand (not-and, or $\bar{\wedge}$), xor (not-biconditional, or $\underline{\vee}$), and nor (not-or, or $\bar{\vee}$).

Other Areas of Mathematics

Generalizing from geometrical symmetry in the previous section, we say that a mathematical object is symmetric with respect to a given mathematical operation, if, when applied to the object, this operation preserves some property of the object. The set of operations that preserve a given property of the object form a group. In general, every kind of structure in mathematics will have its own kind of symmetry. Examples include even and odd functions in calculus; the symmetric group in abstract algebra; symmetric matrices in linear algebra; and the Galòis group in Galois theory. In statistics, it appears as symmetric probability distributions, and as skewness, asymmetry of distributions.

In Science and Nature

In Physics

Symmetry in physics has been generalized to mean invariance—that is, lack of change—under any kind of transformation, for example arbitrary coordinate transformations. This concept has become one of the most powerful tools of theoretical physics, as it has become evident that practically all laws of nature originate in symmetries. In fact, this role inspired the Nobel laureate PW Anderson to write in his widely read 1972 article More is Different that "it is only slightly overstating the case to say that physics is the study of symmetry." See Noether's theorem (which, in greatly simplified form, states that for every continuous mathematical symmetry, there is a corresponding conserved quantity such as energy or momentum; a conserved current, in Noether's original language); and also, Wigner's classification, which says that the symmetries of the laws of physics determine the properties of the particles found in nature.

Important symmetries in physics include continuous symmetries and discrete symmetries of spacetime; internal symmetries of particles; and supersymmetry of physical theories.

Many animals are approximately mirror-symmetric, though internal organs are often arranged asymmetrically.

In Biology

In biology, the notion of symmetry is mostly used explicitly to describe body shapes. Bilateral animals, including humans, are more or less symmetric with respect to the sagittal plane which divides the body into left and right halves. Animals that move in one direction necessarily have upper and lower sides, head and tail ends, and therefore a left and a right. The head becomes specialized with a mouth and sense organs, and the body becomes bilaterally symmetric for the purpose of movement, with symmetrical pairs of muscles and skeletal elements, though internal organs often remain asymmetric.

Plants and sessile (attached) animals such as sea anemones often have radial or rotational symmetry, which suits them because food or threats may arrive from any direction. Fivefold symmetry is found in the echinoderms, the group that includes starfish, sea urchins, and sea lilies.

In biology, the notion of symmetry is also used as in physics, that is to say to describe the properties of the objects studied, including their interactions. A remarkable property of biological evolution is the changes of symmetry corresponding to the appearance of new parts and dynamics.

In Chemistry

Symmetry is important to chemistry because it undergirds essentially all specific interactions between molecules in nature (i.e., via the interaction of natural and human-made chiral molecules with inherently chiral biological systems). The control of the symmetry of molecules produced in modern chemical synthesis contributes to the ability of scientists to offer therapeutic interventions with minimal side effects. A rigorous understanding of symmetry explains fundamental observations in quantum chemistry, and in the applied areas of spectroscopy and crystallography. The theory and application of symmetry to these areas of physical science draws heavily on the mathematical area of group theory.

In Social Interactions

People observe the symmetrical nature, often including asymmetrical balance, of social interactions in a variety of contexts. These include assessments of Reciprocity, empathy, sympathy, apology, dialog, respect, justice, and revenge. Reflective equilibrium is the balance that may be attained through deliberative mutual adjustment among general principles and specific judgments. Symmetrical interactions send the moral message "we are all the same" while asymmetrical interactions may send the message "I am special; better than you." Peer relationships, such as can be governed by the golden rule, are based on symmetry, whereas power relationships are based on asymmetry. Symmetrical relationships can to some degree be maintained by simple (game theory) strategies seen in symmetric games such as tit for tat.

In the Arts

The ceiling of Lotfollah mosque, Isfahan, Iran has 8-fold symmetries.

In Architecture

Seen from the side, the Taj Mahal has bilateral symmetry; from the top (in plan), it has fourfold symmetry.

Symmetry finds its ways into architecture at every scale, from the overall external views of buildings such as Gothic cathedrals and The White House, through the layout of the individual floor plans, and down to the design of individual building elements such as tile mosaics. Islamic buildings such as the Taj Mahal and the Lotfollah mosque make elaborate use of symmetry both in their structure and in their ornamentation. Moorish buildings like the Alhambra are ornamented with complex patterns made using translational and reflection symmetries as well as rotations.

It has been said that only bad architects rely on a "symmetrical layout of blocks, masses and structures";Modernist architecture, starting with International style, relies instead on "wings and balance of masses".

In Pottery and Metal Vessels

Clay pots thrown on a pottery wheel acquire rotational symmetry.

Since the earliest uses of pottery wheels to help shape clay vessels, pottery has had a strong relationship to symmetry. Pottery created using a wheel acquires full rotational symmetry in its cross-section, while allowing substantial freedom of shape in the vertical direction. Upon this inherently symmetrical starting point, potters from ancient times onwards have added patterns that modify the rotational symmetry to achieve visual objectives.

Cast metal vessels lacked the inherent rotational symmetry of wheel-made pottery, but otherwise provided a similar opportunity to decorate their surfaces with patterns pleasing to those who used them. The ancient Chinese, for example, used symmetrical patterns in their bronze castings as early as the 17th century BC. Bronze vessels exhibited both a bilateral main motif and a repetitive translated border design.

In Quilts

Kitchen Kaleidoscope Block

As quilts are made from square blocks (usually 9, 16, or 25 pieces to a block) with each smaller piece usually consisting of fabric triangles, the craft lends itself readily to the application of symmetry.

In Carpets and Rugs

Persian rug with quadrilateral symmetry

A long tradition of the use of symmetry in carpet and rug patterns spans a variety of cultures. American Navajo Indians used bold diagonals and rectangular motifs. Many Oriental rugs have intricate reflected centers and borders that translate a pattern. Not surprisingly, rectangular rugs typically use quadrilateral symmetry—that is, motifs that are reflected across both the horizontal and vertical axes.

In Music

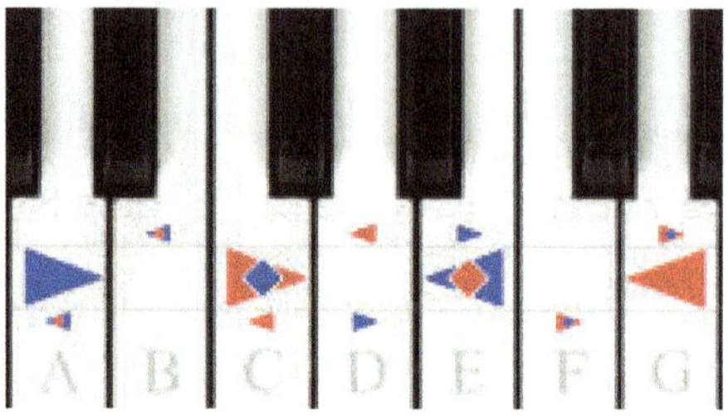

Major and minor triads on the white piano keys are symmetrical to the D. (compare article) (file)

Symmetry is not restricted to the visual arts. Its role in the history of music touches many aspects of the creation and perception of music.

Musical Form

Symmetry has been used as a formal constraint by many composers, such as the arch (swell) form (ABCBA) used by Steve Reich, Béla Bartók, and James Tenney. In classical music, Bach used the symmetry concepts of permutation and invariance.

Pitch Structures

Symmetry is also an important consideration in the formation of scales and chords, traditional or tonal music being made up of non-symmetrical groups of pitches, such as the diatonic scale or the major chord. Symmetrical scales or chords, such as the whole tone scale, augmented chord, or diminished seventh chord (diminished-diminished seventh), are said to lack direction or a sense of forward motion, are ambiguous as to the key or tonal center, and have a less specific diatonic functionality. However, composers such as Alban Berg, Béla Bartók, and George Perle have used axes of symmetry and/or interval cycles in an analogous way to keys or non-tonal tonal centers.

Perle (1992) explains C–E, D–F, and Eb–G, are different instances of the same interval … the other kind of identity. … has to do with axes of symmetry. C–E belongs to a family of symmetrically related dyads as follows:

D	D	**E**	F	F	G	G
D	C	**C**	B	A	A	G

Thus in addition to being part of the interval-4 family, C–E is also a part of the sum-4 family (with C equal to 0).

	2	3	**4**	5	6	7	8
+	2	1	**0**	11	10	9	8
	4	4	4	4	4	4	4

Interval cycles are symmetrical and thus non-diatonic. However, a seven pitch segment of C5 (the cycle of fifths, which are enharmonic with the cycle of fourths) will produce the diatonic major scale. Cyclic tonal progressions in the works of Romantic composers such as Gustav Mahler and Richard Wagner form a link with the cyclic pitch successions in the atonal music of Modernists such as Bartók, Alexander Scriabin, Edgard Varèse, and the Vienna school. At the same time, these progressions signal the end of tonality.

The first extended composition consistently based on symmetrical pitch relations was probably Alban Berg's Quartet, Op. 3 (1910).

Equivalency

Tone rows or pitch class sets which are invariant under retrograde are horizontally symmetrical, under inversion vertically.

In other Arts and Crafts

Celtic knotwork

Symmetries appear in the design of objects of all kinds. Examples include beadwork, furniture, sand paintings, knotwork, masks, and musical instruments. Symmetries are central to the art of M.C. Escher and the many applications of tessellation in art and craft forms such as wallpaper, ceramic tilework, batik, ikat, carpet-making, and many kinds of textile and embroidery patterns.

In Aesthetics

The relationship of symmetry to aesthetics is complex. Humans find bilateral symmetry in faces physically attractive; it indicates health and genetic fitness. Opposed to this is the tendency for excessive symmetry to be perceived as boring or uninteresting. People prefer shapes that have some symmetry, but enough complexity to make them interesting.

In Literature

Symmetry can be found in various forms in literature, a simple example being the palindrome where a brief text reads the same forwards or backwards. Stories may have a symmetrical structure, as in the rise:fall pattern of Beowulf.

REFERENCES

1. Underwood Dudley (1983), "What To Do When the Trisector Comes" (PDF), The Mathematical Intelligencer, 5 (1): 20–25, doi:10.1007/bf03023502
2. Godfried Toussaint, "A new look at Euclid's second proposition," The Mathematical Intelligencer, Vol. 15, No. 3, (1993), pp. 12-24.
3. Bold, Benjamin. Famous Problems of Geometry and How to Solve Them, Dover Publications, 1982 (orig. 1969).
4. Wantzel, P M L (1837). "Recherches sur les moyens de reconnaître si un problème de Géométrie peut se résoudre avec la règle et le compas."(PDF). Journal de Mathématiques Pures et Appliquées. 1. 2: 366–372. Retrieved 3 March 2014.
5. Weisstein, Eric W. "Trigonometry Angles--Pi/17". MathWorld.
6. Stewart, Ian. Galois Theory. p. 75.
7. *Squaring the circle at MacTutor
8. Kazarinoff, Nicholas D. (2003). Ruler and the Round. Mineola, N.Y.: Dover. pp. 29–30. ISBN 0-486-42515-0.
9. Pascal Schreck, Pascal Mathis, Vesna Marinkoviċ, and Predrag Janičiċ. "Wernick's list: A final update", Forum Geometricorum 16, 2016, pp. 69–80. http://forumgeom.fau.edu/FG2016volume16/FG201610.pdf
10. Posamentier, Alfred S., and Lehmann, Ingmar. The Secrets of Triangles, Prometheus Books, 2012.
11. Azad, H., and Laradji, A., "Some impossible constructions in elementary geometry", Mathematical Gazette 88, November 2004, 548–551.
12. Arnon Avron, "On strict strong constructibility with a compass alone", Journal of Geometry (1990) 38: 12.
13. T.L. Heath, "A History of Greek Mathematics, Volume I"
14. P. Hummel, "Solid constructions using ellipses", The Pi Mu Epsilon Journal, 11(8), 429 -- 435 (2003)

15. Gleason, Andrew: "Angle trisection, the heptagon, and the triskaidecagon", Amer. Math. Monthly 95 (1988), no. 3, 185-194.
16. Row, T. Sundara (1966). Geometric Exercises in Paper Folding. New York: Dover.
17. Conway, John H. and Richard Guy: The Book of Numbers
18. A. Baragar, "Constructions using a Twice-Notched Straightedge", The American Mathematical Monthly, 109 (2), 151 -- 164 (2002).
19. E. Benjamin, C. Snyder, "On the construction of the regular hendecagon by marked ruler and compass", Mathematical Proceedings of the Cambridge Philosophical Society, 156 (3), 409 -- 424 (2014).
20. Simon Plouffe (1998). "The Computation of Certain Numbers Using a Ruler and Compass". Journal of Integer Sequences. 1. ISSN 1530-7638.
21. "Curious About Astronomy". Curious.astro.cornell.edu. Archived from the original on 2014-01-11. Retrieved 2014-03-03.
22. "MathWorld: Dimension". Mathworld.wolfram.com. 2014-02-27. Retrieved 2014-03-03.
23. Yau, S-T and Nadis, S.; The Shape of Inner Space, Basic Books, 2010, Chapter 4.
24. Fantechi, Barbara (2001), "Stacks for everybody" (PDF), European Congress of Mathematics Volume I, Progr. Math., 201, Birkhäuser, pp. 349–359
25. Fractal Dimension, Boston University Department of Mathematics and Statistics
26. Bunde, Armin; Havlin, Shlomo, eds. (1991). Fractals and Disordered Systems. Springer.
27. Bunde, Armin; Havlin, Shlomo, eds. (1994). Fractals in Science. Springer.
28. CMS Collaboration, "Search for Microscopic Black Hole Signatures at the Large Hadron Collider" (arxiv.org)
29. Brandenberger, R., Vafa, C., Superstrings in the early universe
30. Scott Watson, Brane Gas Cosmology (pdf).
31. Song, Chaoming; Havlin, Shlomo; Makse, Hernán A. (2005). "Self-similarity of complex networks". Nature. 433 (7024). arXiv:cond-mat/0503078v1 . Bibcode:2005Natur.433..392S. doi:10.1038/nature03248.
32. Daqing, Li; Kosmidis, Kosmas; Bunde, Armin; Havlin, Shlomo (2011). "Dimension of spatially embedded networks". Nature Physics. 7 (6). Bibcode:2011NatPh...7..481D. doi:10.1038/nphys1932.
33. Prolegomena, § 12
34. Banchoff, Thomas F. (1990). "From Flatland to Hypergraphics: Interacting with Higher Dimensions". Interdisciplinary Science Reviews. 15 (4): 364. doi:10.1179/030801890789797239.
35. Newcomb, Simon (1898). "The Philosophy of Hyperspace". Bulletin of the American Mathematical Society. 4 (5): 187. doi:10.1090/S0002-9904-1898-00478-0.

36. Kruger, Runette (2007). "Art in the Fourth Dimension: Giving Form to Form – The Abstract Paintings of Piet Mondrian" (PDF). Spaces of Utopia: an Electronic Journal (5): 11.
37. Pickover, Clifford A. (2009), "Tesseract", The Math Book: From Pythagoras to the 57th Dimension, 250 Milestones in the History of Mathematics, Sterling Publishing Company, Inc., p. 282, ISBN 9781402757969.
38. Lockwood p. ix
39. Heath p. 153
40. Heath p. 160
41. Lockwood p. 132
42. Lockwood p. 129
43. O'Connor, John J.; Robertson, Edmund F., "Spiral of Archimedes", MacTutor History of Mathematics archive, University of St Andrews.
44. "Jordan arc definition at Dictionary.com. Dictionary.com Unabridged. Random House, Inc". Dictionary.reference.com. Retrieved 2012-03-14.
45. Osgood, William F. (January 1903). "A Jordan Curve of Positive Area". Transactions of the American Mathematical Society. American Mathematical Society. 4 (1): 107–112. doi:10.2307/1986455. ISSN 0002-9947. JSTOR 1986455.
46. Liddell and Scott, A Greek–English Lexicon s.v. ρθός
47. Liddell and Scott, A Greek–English Lexicon s.v. γονία
48. Liddell and Scott, A Greek–English Lexicon s.v. ρθογώνιον
49. Oxford English Dictionary, Third Edition, September 2004, s.v. orthogonal
50. J.A. Wheeler; C. Misner; K.S. Thorne (1973). Gravitation. W.H. Freeman & Co. p. 58. ISBN 0-7167-0344-0.
51. Wolfram MathWorld.
52. Bourbaki, ch. II §2.4, Algebra I, p. 234
53. Trefethen, Lloyd N. & Bau, David (1997). Numerical linear algebra. SIAM. p. 13. ISBN 978-0-89871-361-9.
54. R. Penrose (2007). The Road to Reality. Vintage books. pp. 417–419. ISBN 0-679-77631-1.
55. Michael L. Scott, Programming Language Pragmatics, p. 228
56. 1968, Adriaan van Wijngaarden et al., Revised Report on the Algorithmic Language ALGOL 68, section 0.1.2, Orthogonal design
57. Null, Linda & Lobur, Julia (2006). The essentials of computer organization and architecture (2nd ed.). Jones & Bartlett Learning. p. 257. ISBN 978-0-7637-3769-6.
58. Athanasios Papoulis; S. Unnikrishna Pillai (2002). Probability, Random Variables and Stochastic Processes. McGraw-Hill. p. 211. ISBN 0-07-366011-6.
59. Hedayat, A.; et al. (1999). Orthogonal arrays: theory and applications. Springer. p. 168. ISBN 978-0-387-98766-8.

60. Karver, Mark R.; Hilderbrand, Scott A. (2012). Bioorthogonal Reaction Pairs Enable Simultaneous, Selective, Multi-Target Imaging. Angewandte Chemie International Edition. 51 (4): 920–2. doi:10.1002/anie.201104389. PMC 3304098 . PMID 22162316.
61. chessvariants.org chess glossary.
62. For an illustration, see YouTube.
63. Although this postulate only refers to when lines meet, it is needed to prove the uniqueness of parallel lines in the sense of Playfair's axiom.
64. Mathematical Operators – Unicode Consortium (PDF). Retrieved 2013-04-21.
65. Wylie, Jr. 1964, pp. 92—94
66. Heath 1956, pp. 190–194
67. Richards 1988, Chap. 4: Euclid and the English Schoolchild. pp. 161–200
68. Carroll, Lewis (2009) 1879, Euclid and His Modern Rivals, Barnes & Noble, ISBN 978-1-4351-2348-9
69. Wilson 1868
70. Einführung in die Grundlagen der Geometrie, I, p. 5
71. Heath 1956, p. 194
72. Richards 1988, pp. 180–184
73. Heath 1956, p. 194
74. Only the third is a straightedge and compass construction, the first two are infinitary processes (they require an infinite number of steps.)
75. H. S. M. Coxeter (1961) Introduction to Geometry, p 192, John Wiley & Sons
76. Wanda Szmielew (1983) From Affine to Euclidean Geometry, p 17, D. Reidel ISBN 90-277-1243-3
77. Andy Liu (2011) Is parallelism an equivalence relation?, The College Mathematics Journal 42(5):372
78. Weisstein, Eric W. Vertex. MathWorld.
79. Heath, Thomas L. (1956). The Thirteen Books of Euclid's Elements (2nd ed. Facsimile. Original publication: Cambridge University Press, 1925 ed.). New York: Dover Publications.
80. (3 vols.): ISBN 0-486-60088-2 (vol. 1), ISBN 0-486-60089-0 (vol. 2), ISBN 0-486-60090-4 (vol. 3).
81. Jing, Lanru; Stephansson, Ove (2007). Fundamentals of Discrete Element Methods for Rock Engineering: Theory and Applications. Elsevier Science.
82. Peter McMullen, Egon Schulte, Abstract Regular Polytopes, Cambridge University Press, 2002. ISBN 0-521-81496-0 (Page 29)
83. Bobenko, Alexander I.; Schröder, Peter; Sullivan, John M.; Ziegler, Günter M. (2008). Discrete differential geometry. Birkhäuser Verlag AG. ISBN 978-3-7643-8620-7.

84. M.V. Jaric, ed, Introduction to the Mathematics of Quasicrystals (Aperiodicity and Order, Vol 2) ISBN 0-12-040602-0, Academic Press, 1989.
85. Devadoss, Satyan; O'Rourke, Joseph (2011). Discrete and Computational Geometry. Princeton University Press. ISBN 978-0-691-14553-2.
86. Meisters, G. H. (1975), Polygons have ears, The American Mathematical Monthly, 82: 648–651, doi:10.2307/2319703, MR 0367792.
87. Christen, Martin. Clockworkcoders Tutorials: Vertex Attributes. Khronos Group. Retrieved 26 January 2009.
88. Clapham, C.; Nicholson, J. (2009). Oxford Concise Dictionary of Mathematics, Congruent Figures (PDF). Addison-Wesley. p. 167. Retrieved September 2013. Check date values in: |access-date= (help)
89. Congruence. Math Open Reference. 2009. Retrieved September 2013. Check date values in: |access-date= (help)
90. Parr, H. E. (1970). Revision Course in School mathematics. Mathematics Textbooks Second Edition. G Bell and Sons Ltd. ISBN 0-7135-1717-4.
91. Cornel, Antonio (2002). Geometry for Secondary Schools. Mathematics Textbooks Second Edition. Bookmark Inc. ISBN 971-569-441-1.
92. Alexander Borisov, Mark Dickinson, and Stuart Hastings, A congruence problem for polyhedra, American Mathematical Monthly 117, March 2010, pp. 232-249.
93. Alexa Creech, A congruence problem http://146.163.152.131/teaching/projects/creech_final.pdf Archived November 11, 2013, at the Wayback Machine.
94. Michael Bolin, Exploration of Spherical Geometry, September 9, 2003, pp.6-7. http://math.iit.edu/~mccomic/420/notes/Bolin_spherical.pdf
95. Hollyer, L., http://www.uh.edu/~hollyer/Module6/m6ppt/sld089.htm
96. symmetry. Online Etymology Dictionary.
97. Zee, A. (2007). Fearful Symmetry. Princeton, N.J.: Princeton University Press. ISBN 978-0-691-13482-6.
98. Symmetry and the Beautiful Universe, Christopher T. Hill and Leon M. Lederman, Prometheus Books (2005)
99. Mainzer, Klaus (2005). Symmetry And Complexity: The Spirit and Beauty of Nonlinear Science. World Scientific. ISBN 981-256-192-7.
100. E. H. Lockwood, R. H. Macmillan, Geometric Symmetry, London: Cambridge Press, 1978
101. Weyl, Hermann (1982) [1952]. Symmetry. Princeton: Princeton University Press. ISBN 0-691-02374-3.
102. Singer, David A. (1998). Geometry: Plane and Fancy. Springer Science & Business Media.
103. Stenger, Victor J. (2000) and Mahou Shiro (2007). Timeless Reality. Prometheus Books. Especially chapter 12. Nontechnical.
104. Bottema, O, and B. Roth, Theoretical Kinematics, Dover Publications (September 1990)

105. Tian Yu Cao Conceptual Foundations of Quantum Field Theory Cambridge University Press p.154-155

106. Gouyet, Jean-François (1996). Physics and fractal structures. Paris/New York: Masson Springer. ISBN 978-0-387-94153-0.

107. Josiah Royce, Ignas K. Skrupskelis (2005) The Basic Writings of Josiah Royce: Logic, loyalty, and community (Google eBook) Fordham Univ Press, p. 790

108. Christopher G. Morris (1992) Academic Press Dictionary of Science and Technology Gulf Professional Publishing

109. Costa, Giovanni; Fogli, Gianluigi (2012). Symmetries and Group Theory in Particle Physics: An Introduction to Space-Time and Internal Symmetries. Springer Science & Business Media. p. 112.

110. Anderson, P.W. (1972). More is Different (PDF). Science. 177 (4047): 393–396. Bibcode:1972Sci...177..393A. doi:10.1126/science.177.4047.393. PMID 17796623.

111. Kosmann-Schwarzbach, Yvette (2010). The Noether theorems: Invariance and conservation laws in the twentieth century. Sources and Studies in the History of Mathematics and Physical Sciences. Springer-Verlag. ISBN 978-0-387-87867-6.

112. Wigner, E. P. (1939), On unitary representations of the inhomogeneous Lorentz group, Annals of Mathematics, 40 (1): 149–204, Bibcode:1939AnMat..40..149W, doi:10.2307/1968551, MR 1503456.

113. Valentine, James W. Bilateria. AccessScience. Retrieved 29 May 2013.

114. Hickman, Cleveland P.; Roberts, Larry S.; Larson, Allan (2002). Animal Diversity (Third Edition) (PDF). Chapter 8: Acoelomate Bilateral Animals. McGraw-Hill. p. 139. Retrieved October 25, 2012.

115. Stewart, Ian (2001). What Shape is a Snowflake? Magical Numbers in Nature. Weidenfeld & Nicolson. pp. 64–65.

116. Longo, Giuseppe; Montévil, Maël (2016). Perspectives on Organisms: Biological time, Symmetries and Singularities. Springer. ISBN 978-3-662-51229-6.

117. Montévil, Maël; Mossio, Matteo; Pocheville, Arnaud; Longo, Giuseppe (2016). Theoretical principles for biology: Variation. Progress in Biophysics and Molecular Biology. From the Century of the Genome to the Century of the Organism: New Theoretical Approaches. 122 (1): 36–50. doi:10.1016/j.pbiomolbio.2016.08.005.

118. Lowe, John P; Peterson, Kirk (2005). Quantum Chemistry (Third ed.). Academic Press. ISBN 0-12-457551-X.

119. Daniels, Norman (2003-04-28). Reflective Equilibrium. Stanford Encyclopedia of Philosophy.

120. Emotional Competency: Symmetry

121. Lutus, P. (2008). The Symmetry Principle. Retrieved 28 September 2015.

122. Williams: Symmetry in Architecture. Members.tripod.com (1998-12-31). Retrieved on 2013-04-16.

123. Aslaksen: Mathematics in Art and Architecture. Math.nus.edu.sg. Retrieved on 2013-04-16.
124. Derry, Gregory N. (2002). What Science Is and How It Works. Princeton University Press. pp. 269–. ISBN 978-1-4008-2311-6.
125. Dunlap, David W. (31 July 2009). Behind the Scenes: Edgar Martins Speaks. New York Times. Retrieved 11 November 2014. "My starting point for this construction was a simple statement which I once read (and which does not necessarily reflect my personal views): 'Only a bad architect relies on symmetry; instead of symmetrical layout of blocks, masses and structures, Modernist architecture relies on wings and balance of masses.'
126. The Art of Chinese Bronzes. Chinavoc (2007-11-19). Retrieved on 2013-04-16.
127. Quate: Exploring Geometry Through Quilts. Its.guilford.k12.nc.us. Retrieved on 2013-04-16.
128. Marla Mallett Textiles & Tribal Oriental Rugs. The Metropolitan Museum of Art, New York.
129. Dilucchio: Navajo Rugs. Navajocentral.org (2003-10-26). Retrieved on 2013-04-16.
130. see (Fugue No. 21, pdf or Shockwave)
131. Perle, George (1992). Symmetry, the twelve-tone scale, and tonality. Contemporary Music Review. 6 (2): 81–96. doi:10.1080/07494469200640151.
132. Perle, George (1990). The Listening Composer. University of California Press.
133. Cucker, Felix (2013). Manifold Mirrors: The Crossing Paths of the Arts and Mathematics. Cambridge University Press. pp. 77–78, 83, 89, 103. ISBN 978-0-521-72876-8.
134. Grammer, K.; Thornhill, R. (1994). Human (Homo sapiens) facial attractiveness and sexual selection: the role of symmetry and averageness. Journal of Comparative Psychology. Washington, D.C. 108 (3): 233–42. doi:10.1037/0735-7036.108.3.233.
135. Rhodes, Gillian; Zebrowitz, Leslie, A. (2002). Facial Attractiveness - Evolutionary, Cognitive, and Social Perspectives. Ablex. ISBN 1-56750-636-4.
136. Jones, B. C., Little, A. C., Tiddeman, B. P., Burt, D. M., & Perrett, D. I. (2001). Facial symmetry and judgements of apparent health Support for a "' good genes '" explanation of the attractiveness – symmetry relationship, 22, 417–429.
137. Arnheim, Rudolf (1969). Visual Thinking. University of California Press.

CHAPTER 3

Classical Branches

Chapter Contents

INTRODUCTION

Algebraic geometry is a branch of mathematics, classically studying zeros of multivariate polynomials. Modern algebraic geometry is based on the use of abstract algebraic techniques, mainly from commutative algebra, for solving geometrical problems about these sets of zeros.

The fundamental objects of study in algebraic geometry are algebraic varieties, which are geometric manifestations of solutions of systems of polynomial equations. Examples of the most studied classes of algebraic varieties are: plane algebraic curves, which include lines, circles, parabolas, ellipses, hyperbolas, cubic curves like elliptic curves and quartic curves like lemniscates, and Cassini ovals. A point of the plane belongs to an algebraic curve if its coordinates satisfy a given polynomial equation. Basic questions involve the study of the points of special interest like the singular points, the inflection points and the points at infinity. More advanced questions involve the topology of the curve and relations between the curves given by different equations.

Algebraic geometry occupies a central place in modern mathematics and has multiple conceptual connections with such diverse fields as complex analysis, topology and number theory. Initially a study of systems of polynomial equations in several variables, the subject of algebraic geometry starts where equation solving leaves off, and it becomes even more important to understand the intrinsic properties of the totality of solutions of a system of equations, than to find a specific solution; this leads into some of the deepest areas in all of mathematics, both conceptually and in terms of technique.

In the 20th century, algebraic geometry split into several subareas.

- The mainstream of algebraic geometry is devoted to the study of the complex points of the algebraic varieties and more generally to the points with coordinates in an algebraically closed field.
- The study of the points of an algebraic variety with coordinates in the field of the rational numbers or in a number field became arithmetic geometry (or more classically Diophantine geometry), a subfield of algebraic number theory.
- The study of the real points of an algebraic variety is the subject of real algebraic geometry.
- A large part of singularity theory is devoted to the singularities of algebraic varieties.
- With the rise of the computers, a computational algebraic geometry area has emerged, which lies at the intersection of algebraic geometry and computer algebra. It consists essentially in developing algorithms and software for studying and finding the properties of explicitly given algebraic varieties.

Much of the development of the mainstream of algebraic geometry in the 20th century occurred within an abstract algebraic framework, with increasing emphasis being placed on intrinsic properties of algebraic varieties not dependent on any particular way of embedding the variety in an ambient coordinate space; this parallels developments in topology, differential and complex geometry. One key achievement of this abstract algebraic geometry is Grothendieck›s scheme theory which allows one to use sheaf theory to study algebraic varieties in a way which is very similar to its use in the study of differential and analytic manifolds. This is obtained by extending the notion of point: In classical algebraic geometry, a point of an affine variety may be identified, through Hilbert›s Nullstellensatz, with a maximal ideal of the coordinate ring, while the points of the corresponding affine scheme are all prime ideals of this ring. This means that a point of such a scheme may be either a usual point or a subvariety. This approach also enables a unification of the language and the tools of classical algebraic geometry, mainly concerned with complex points, and of algebraic number theory. Wiles›s proof of the longstanding conjecture called Fermat›s last theorem is an example of the power of this approach.

This Togliatti surface is an algebraic surface of degree five. The picture represents a portion of its real locus.

Zeros of Simultaneous Polynomials

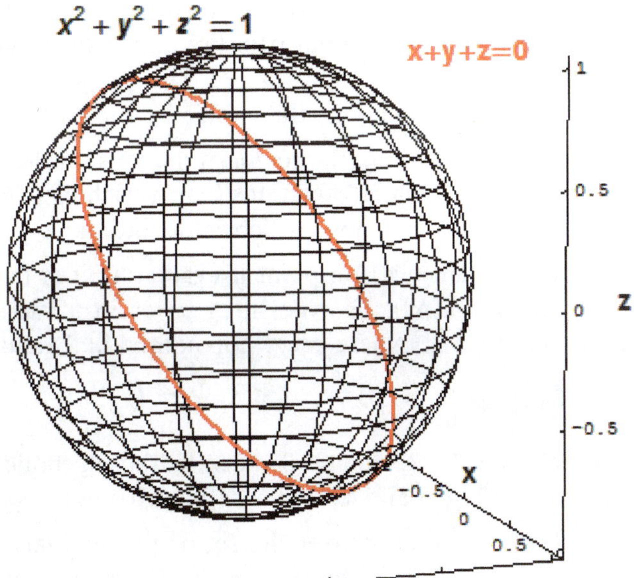

Sphere and slanted circle

In classical algebraic geometry, the main objects of interest are the vanishing sets of collections of polynomials, meaning the set of all points that simultaneously satisfy one or more polynomial equations. For instance, the two-dimensional sphere in three-dimensional Euclidean space $\mathbf{R}^3$ could be defined as the set of all points (x,y,z) with

$$x^2 + y^2 + z^2 - 1 = 0.$$

A "slanted" circle in $\mathbf{R}^3$ can be defined as the set of all points (x,y,z) which satisfy the two polynomial equations

$$x^2 + y^2 + z^2 - 1 = 0,$$
$$x + y + z = 0.$$

Affine Varieties

First we start with a field k. In classical algebraic geometry, this field was always the complex numbers **C**, but many of the same results are true if we assume only that k is algebraically closed. We consider the affine space of dimension n over k, denoted $\mathbf{A}^n(k)$ (or more simply $\mathbf{A}^n$, when k is clear from the context). When one fixes a coordinates system, one may identify $\mathbf{A}^n(k)$ with k^n. The purpose of not working with k^n is to emphasize that one "forgets" the vector space structure that k^n carries.

A function $f : \mathbf{A}^n \rightarrow \mathbf{A}^1$ is said to be polynomial (or regular) if it can be written as a polynomial, that is, if there is a polynomial p in $k[x_1,...,x_n]$ such that $f(M) = p(t_1,...,t_n)$ for every point M with coordinates $(t_1,...,t_n)$ in $\mathbf{A}^n$. The property of a function to be polynomial (or regular) does not depend on the choice of a coordinate system in $\mathbf{A}^n$.

When a coordinate system is chosen, the regular functions on the affine n-space may be identified with the ring of polynomial functions in n variables over k. Therefore, the set of the regular functions on $\mathbf{A}^n$ is a ring, which is denoted $k[\mathbf{A}^n]$.

We say that a polynomial vanishes at a point if evaluating it at that point gives zero. Let S be a set of polynomials in $k[\mathbf{A}^n]$. The vanishing set of S (or vanishing locus or zero set) is the set V(S) of all points in $\mathbf{A}^n$ where every polynomial in S vanishes. Symbolically,

$$V(S) = \{(t_1, \ldots, t_n) | \forall p \in S, p(t_1, \ldots, t_n) = 0\}.$$

A subset of $\mathbf{A}^n$ which is V(S), for some S, is called an algebraic set. The V stands for variety (a specific type of algebraic set to be defined below).

Given a subset U of $\mathbf{A}^n$, can one recover the set of polynomials which generate it? If U is any subset of $\mathbf{A}^n$, define I(U) to be the set of all polynomials whose vanishing set contains U. The I stands for ideal: if two polynomials f and g both vanish on U, then f+g vanishes on U, and if h is any polynomial, then hf vanishes on U, so I(U) is always an ideal of the polynomial ring $k[\mathbf{A}^n]$.

Two natural questions to ask are:

- Given a subset U of $\mathbf{A}^n$, when is U = V(I(U))?

- Given a set S of polynomials, when is S = I(V(S))?

The answer to the first question is provided by introducing the Zariski topology, a topology on $\mathbf{A}^n$ whose closed sets are the algebraic sets, and which directly reflects the algebraic structure of k[$\mathbf{A}^n$]. Then U = V(I(U)) if and only if U is an algebraic set or equivalently a Zariski-closed set. The answer to the second question is given by Hilbert's Nullstellensatz. In one of its forms, it says that I(V(S)) is the radical of the ideal generated by S. In more abstract language, there is a Galois connection, giving rise to two closure operators; they can be identified, and naturally play a basic role in the theory; the example is elaborated at Galois connection.

For various reasons we may not always want to work with the entire ideal corresponding to an algebraic set U. Hilbert's basis theorem implies that ideals in k[$\mathbf{A}^n$] are always finitely generated.

An algebraic set is called irreducible if it cannot be written as the union of two smaller algebraic sets. Any algebraic set is a finite union of irreducible algebraic sets and this decomposition is unique. Thus its elements are called the irreducible components of the algebraic set. An irreducible algebraic set is also called a variety. It turns out that an algebraic set is a variety if and only if it may be defined as the vanishing set of a prime ideal of the polynomial ring.

Some authors do not make a clear distinction between algebraic sets and varieties and use irreducible variety to make the distinction when needed.

Regular Functions

Just as continuous functions are the natural maps on topological spaces and smooth functions are the natural maps on differentiable manifolds, there is a natural class of functions on an algebraic set, called regular functions or polynomial functions. A regular function on an algebraic set V contained in $\mathbf{A}^n$ is the restriction to V of a regular function on $\mathbf{A}^n$. For an algebraic set defined on the field of the complex numbers, the regular functions are smooth and even analytic.

It may seem unnaturally restrictive to require that a regular function always extend to the ambient space, but it is very similar to the situation in a normal topological space, where the Tietze extension theorem guarantees that a continuous function on a closed subset always extends to the ambient topological space.

Just as with the regular functions on affine space, the regular functions on V form a ring, which we denote by k[V]. This ring is called the coordinate ring of V.

Since regular functions on V come from regular functions on $\mathbf{A}^n$, there is a relationship between the coordinate rings. Specifically, if a regular function on V is the restriction of two functions f and g in k[$\mathbf{A}^n$], then f − g is a polynomial function which is null on V and thus belongs to I(V). Thus k[V] may be identified with k[$\mathbf{A}^n$]/I(V).

Morphism of Affine Varieties

Using regular functions from an affine variety to $\mathbf{A}^1$, we can define regular maps from one affine variety to another. First we will define a regular map from a variety into affine space: Let V be a variety contained in $\mathbf{A}^n$. Choose m regular functions on V, and call them $f_1, ..., f_m$. We define a regular map f from V to $\mathbf{A}^m$ by letting $f = (f_1, ..., f_m)$. In other words, each f_i determines one coordinate of the range of f.

If V' is a variety contained in $\mathbf{A}^m$, we say that f is a regular map from V to V' if the range of f is contained in V'.

The definition of the regular maps apply also to algebraic sets. The regular maps are also called morphisms, as they make the collection of all affine algebraic sets into a category, where the objects are the affine algebraic sets and the morphisms are the regular maps. The affine varieties is a subcategory of the category of the algebraic sets.

Given a regular map g from V to V' and a regular function f of k[V'], then $f \circ g \in k[V]$. The map $f \rightarrow f \circ g$ is a ring homomorphism from k[V'] to k[V]. Conversely, every ring homomorphism from k[V'] to k[V] defines a regular map from V to V'. This defines an equivalence of categories between the category of algebraic sets and the opposite category of the finitely generated reduced k-algebras. This equivalence is one of the starting points of scheme theory.

Rational Function and Birational Equivalence

Contrarily to the preceding ones, this section concerns only varieties and not algebraic sets. On the other hand, the definitions extend naturally to projective varieties (next section), as an affine variety and its projective completion have the same field of functions.

If V is an affine variety, its coordinate ring is an integral domain and has thus a field of fractions which is denoted k(V) and called the field of the rational functions on V or, shortly, the function field of V. Its elements are the restrictions to V of the rational functions over the affine space containing V. The domain of a rational function f is not V but the complement of the subvariety (a hypersurface) where the denominator of f vanishes.

Like for regular maps, one may define a rational map from a variety V to a variety V'. Like for the regular maps, the rational maps from V to V' may be identified to the field homomorphisms from k(V') to k(V).

Two affine varieties are birationally equivalent if there are two rational functions between them which are inverse one to the other in the regions where both are defined. Equivalently, they are birationally equivalent if their function fields are isomorphic.

An affine variety is a rational variety if it is birationally equivalent to an affine space. This means that the variety admits a rational parameterization. For example, the circle of equation $x^2 + y^2 - 1 = 0$ is a rational curve, as it has the parameterization

$$x = \frac{2t}{1+t^2}$$
$$y = \frac{1-t^2}{1+t^2},$$

which may also be viewed as a rational map from the line to the circle.

The problem of resolution of singularities is to know if every algebraic variety is birationally equivalent to a variety whose projective completion is nonsingular (see also smooth completion). It has been positively solved in characteristic 0 by Heisuke Hironaka in 1964 and is yet unsolved in finite characteristic.

Projective Variety

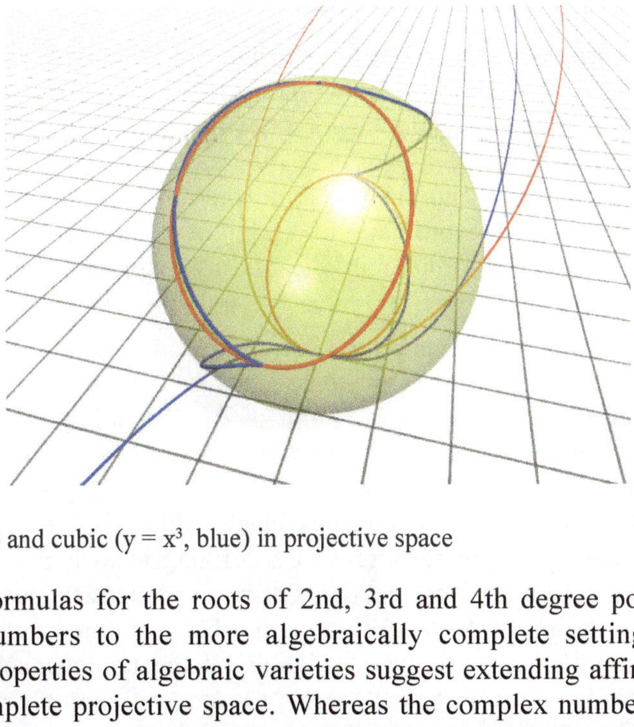

parabola ($y = x^2$, red) and cubic ($y = x^3$, blue) in projective space

Just as the formulas for the roots of 2nd, 3rd and 4th degree polynomials suggest extending real numbers to the more algebraically complete setting of the complex numbers, many properties of algebraic varieties suggest extending affine space to a more geometrically complete projective space. Whereas the complex numbers are obtained by adding the number i, a root of the polynomial x^2 + 1, projective space is obtained by adding in appropriate points "at infinity", points where parallel lines may meet.

To see how this might come about, consider the variety $V(y - x^2)$. If we draw it, we get a parabola. As x goes to positive infinity, the slope of the line from the origin to the point (x, x^2) also goes to positive infinity. As x goes to negative infinity, the slope of the same line goes to negative infinity.

Compare this to the variety $V(y - x^3)$. This is a cubic curve. As x goes to positive infinity, the slope of the line from the origin to the point (x, x^3) goes to positive infinity just as before. But unlike before, as x goes to negative infinity, the slope of the same line goes to positive infinity as well; the exact opposite of the parabola. So the behavior "at infinity" of $V(y - x^3)$ is different from the behavior "at infinity" of $V(y - x^2)$.

The consideration of the projective completion of the two curves, which is their prolongation "at infinity" in the projective plane, allows to quantify this difference: the point at infinity of the parabola is a regular point, whose tangent is the line at infinity, while the point at infinity of the cubic curve is a cusp. Also, both curves are rational, as they are parameterized by x, and Riemann-Roch theorem implies that the cubic curve must have a singularity, which must be at infinity, as all its points in the affine space are regular.

Thus many of the properties of algebraic varieties, including birational equivalence and all the topological properties, depend on the behavior "at infinity" and so it is natural to study the varieties in projective space. Furthermore, the introduction of projective techniques made many theorems in algebraic geometry simpler and sharper: For example, Bézout's theorem on the number of intersection points between two varieties can be stated in its sharpest form only in projective space. For these reasons, projective space plays a fundamental role in algebraic geometry.

Nowadays, the projective space $\mathbf{P}^n$ of dimension n is usually defined as the set of the lines passing through a point, considered as the origin, in the affine space of dimension n+1, or equivalently to the set of the vector lines in a vector space of dimension n+1. When a coordinate system has been chosen in the space of dimension n+1, all the points of a line have the same set of coordinates, up to the multiplication by an element of k. This defines the homogeneous coordinates of a point of $\mathbf{P}^n$ as a sequence of n+1 elements of the base field k, defined up to the multiplication by a nonzero element of k (the same for the whole sequence).

Given a polynomial in n+1 variables, it vanishes at all the point of a line passing through the origin if and only if it is homogeneous. In this case, one says that the polynomial vanishes at the corresponding point of $\mathbf{P}^n$. This allows to define a projective algebraic set in $\mathbf{P}^n$ as the set $V(f_1, ..., f_k)$ where a finite set of homogeneous polynomials $\{f_1, ..., f_k\}$ vanishes. Like for affine algebraic sets, there is a bijection between the projective algebraic sets and the reduced homogeneous ideals which define them. The projective varieties are the projective algebraic sets whose defining ideal is prime. In other words, a projective variety is a projective algebraic set, whose homogeneous coordinate ring is an integral domain, the projective coordinates ring being defined as the quotient of the graded ring or the polynomials in n+1 variables by the homogeneous (reduced) ideal defining the variety. Every projective algebraic set may be uniquely decomposed into a finite union of projective varieties.

The only regular functions which may be defined properly on a projective variety are the constant functions. Thus this notion is not used in projective situations. On the other hand, the field of the rational functions or function field is a useful notion, which, similarly as in the affine case, is defined as the set of the quotients of two homogeneous elements of the same degree in the homogeneous coordinate ring.

Real Algebraic Geometry

The real algebraic geometry is the study of the real points of the algebraic geometry. The fact that the field of the reals number is an ordered field should not be ignored in such a study. For example, the curve of equation $x^2 + y^2 - a = 0$ is a circle if $a > 0$, but does not have any real point if $a < 0$. It follows that real algebraic geometry is not only the study of the real algebraic varieties, but has been generalized to the study of the semi-algebraic sets, which are the solutions of systems of polynomial equations and polynomial inequalities. For example, a branch of the hyperbola of equation $xy - 1 = 0$ is not an algebraic variety, but is a semi-algebraic set defined by $xy - 1 = 0$ and $x > 0$ or by $xy - 1 = 0$ and $x + y > 0$.

One of the challenging problems of real algebraic geometry is the unsolved Hilbert's sixteenth problem: Decide which respective positions are possible for the ovals of a nonsingular plane curve of degree 8.

Computational Algebraic Geometry

One may date the origin of computational algebraic geometry to meeting EUROSAM'79 (International Symposium on Symbolic and Algebraic Manipulation) held at Marseille, France in June 1979. At this meeting,

- Dennis S. Arnon showed that George E. Collins's Cylindrical algebraic decomposition (CAD) allows the computation of the topology of semi-algebraic sets,
- Bruno Buchberger presented the Gröbner bases and his algorithm to compute them,
- Daniel Lazard presented a new algorithm for solving systems of homogeneous polynomial equations with a computational complexity which is essentially polynomial in the expected number of solutions and thus simply exponential in the number of the unknowns. This algorithm is strongly related with Macaulay's multivariate resultant.

Since then, most results in this area are related to one or several of these items either by using or improving one of these algorithms, or by finding algorithms whose complexity is simply exponential in the number of the variables

A body of mathematical theory complementary to symbolic methods called numerical algebraic geometry has been developed over the last several decades. The main computational method is homotopy continuation. This supports, for example, a model of floating point computation for solving problems of algebraic geometry.

Gröbner Basis

A Gröbner basis is a system of generators of a polynomial ideal whose computation allows the deduction of many properties of the affine algebraic variety defined by the ideal.

Given an ideal I defining an algebraic set V:

- V is empty (over an algebraically closed extension of the basis field), if and only if the Gröbner basis for any monomial ordering is reduced to {1}.
- By means of the Hilbert series one may compute the dimension and the degree of V from any Gröbner basis of I for a monomial ordering refining the total degree.
- If the dimension of V is 0, one may compute the points (finite in number) of V from any Gröbner basis of I (see systems of polynomial equations).
- A Gröbner basis computation allows to remove from V all irreducible components which are contained in a given hyper surface.
- A Gröbner basis computation allows to compute the Zariski closure of the image of V by the projection on the k first coordinates, and the subset of the image where the projection is not proper.
- More generally Gröbner basis computations allows to compute the Zariski closure of the image and the critical points of a rational function of V into another affine variety.

Gröbner basis computations do not allow to compute directly the primary decomposition of I nor the prime ideals defining the irreducible components of V, but most algorithms for this involve Gröbner basis computation. The algorithms which are not based on Gröbner bases use regular chains but may need Gröbner bases in some exceptional situations.

Gröbner base are deemed to be difficult to compute. In fact they may contain, in the worst case, polynomials whose degree is doubly exponential in the number of variables and a number of polynomials which is also doubly exponential. However, this is only a worst case complexity, and the complexity bound of Lazard's algorithm of 1979 may frequently apply. Faugère's F4 and F5 algorithms realize this complexity, as F5 algorithm may be viewed as an improvement of Lazard's 1979 algorithm. It follows that the best implementations allow to compute almost routinely with algebraic sets of degree more than 100. This means that, presently, the difficulty of computing a Gröbner basis is strongly related to the intrinsic difficulty of the problem.

Cylindrical Algebraic Decomposition (CAD)

CAD is an algorithm which had been introduced in 1973 by G. Collins to implement with an acceptable complexity the Tarski–Seidenberg theorem on quantifier elimination over the real numbers.

This theorem concerns the formulas of the first-order logic whose atomic formulas are polynomial equalities or inequalities between polynomials with real coefficients. These formulas are thus the formulas which may be constructed from the atomic formulas by the logical operators and (∧), or (∨), not (¬), for all (∀) and exists (∃). Tarski's theorem asserts that, from such a formula, one may compute an equivalent formula without quantifier (∀, ∃).

The complexity of CAD is doubly exponential in the number of variables. This means that CAD allow, in theory, to solve every problem of real algebraic geometry which may be expressed by such a formula, that is almost every problem concerning explicitly given varieties and semi-algebraic sets.

While Gröbner basis computation has doubly exponential complexity only in rare cases, CAD has almost always this high complexity. This implies that, unless if most polynomials appearing in the input are linear, it may not solve problems with more than four variables.

Since 1973, most of the research on this subject is devoted either to improve CAD or to find alternate algorithms in special cases of general interest.

As an example of the state of art, there are efficient algorithms to find at least a point in every connected component of a semi-algebraic set, and thus to test if a semi-algebraic set is empty. On the other hand, CAD is yet, in practice, the best algorithm to count the number of connected components.

Asymptotic Complexity Vs. Practical Efficiency

The basic general algorithms of computational geometry have a double exponential worst case complexity. More precisely, if d is the maximal degree of the input polynomials and n the number of variables, their complexity is at most $d^{2^{cn}}$ for some constant c, and, for some inputs, the complexity is at least $d^{2^{c'n}}$ for another constant c′.

During the last 20 years of 20th century, various algorithms have been introduced to solve specific subproblems with a better complexity. Most of these algorithms have a complexity $d^{O(n^2)}$.

Among these algorithms which solve a sub problem of the problems solved by Gröbner bases, one may cite testing if an affine variety is empty and solving nonhomogeneous polynomial systems which have a finite number of solutions. Such algorithms are rarely implemented because, on most entries Faugère's F4 and F5 algorithms have a better practical efficiency and probably a similar or better complexity (probably because the evaluation of the complexity of Gröbner basis algorithms on a particular class of entries is a difficult task which has been done only in a few special cases).

The main algorithms of real algebraic geometry which solve a problem solved by CAD are related to the topology of semi-algebraic sets. One may cite counting the number of connected components, testing if two points are in the same components or computing a Whitney stratification of a real algebraic set. They have a complexity of $d^{O(n^2)}$,

but the constant involved by O notation is so high that using them to solve any nontrivial problem effectively solved by CAD, is impossible even if one could use all the existing computing power in the world. Therefore, these algorithms have never been implemented and this is an active research area to search for algorithms with have together a good asymptotic complexity and a good practical efficiency.

Abstract Modern Viewpoint

The modern approaches to algebraic geometry redefine and effectively extend the range of basic objects in various levels of generality to schemes, formal schemes, ind-schemes, algebraic spaces, algebraic stacks and so on. The need for this arises already from the useful ideas within theory of varieties, e.g. the formal functions of Zariski can be accommodated by introducing nilpotent elements in structure rings; considering spaces of loops and arcs, constructing quotients by group actions and developing formal grounds for natural intersection theory and deformation theory lead to some of the further extensions.

Most remarkably, in late 1950s, algebraic varieties were subsumed into Alexander Grothendieck's concept of a scheme. Their local objects are affine schemes or prime spectra which are locally ringed spaces which form a category which is antiequivalent to the category of commutative unital rings, extending the duality between the category of affine algebraic varieties over a field k, and the category of finitely generated reduced k-algebras. The gluing is along Zariski topology; one can glue within the category of locally ringed spaces, but also, using the Yoneda embedding, within the more abstract category of presheaves of sets over the category of affine schemes. The Zariski topology in the set theoretic sense is then replaced by a Grothendieck topology. Grothendieck introduced Grothendieck topologies having in mind more exotic but geometrically finer and more sensitive examples than the crude Zariski topology, namely the étale topology, and the two flat Grothendieck topologies: fppf and fpqc; nowadays some other examples became prominent including Nisnevich topology. Sheaves can be furthermore generalized to stacks in the sense of Grothendieck, usually with some additional representability conditions leading to Artin stacks and, even finer, Deligne-Mumford stacks, both often called algebraic stacks.

Sometimes other algebraic sites replace the category of affine schemes. For example, Nikolai Durov has introduced commutative algebraic monads as a generalization of local objects in a generalized algebraic geometry. Versions of a tropical geometry, of an absolute geometry over a field of one element and an algebraic analogue of Arakelov's geometry were realized in this setup.

Another formal generalization is possible to Universal algebraic geometry in which every variety of algebras has its own algebraic geometry. The term variety of algebras should not be confused with algebraic variety.

The language of schemes, stacks and generalizations has proved to be a valuable way of dealing with geometric concepts and became cornerstones of modern algebraic geometry.

Algebraic stacks can be further generalized and for many practical questions like deformation theory and intersection theory, this is often the most natural approach. One can extend the Grothendieck site of affine schemes to a higher categorical site of derived affine schemes, by replacing the commutative rings with an infinity category of differential graded commutative algebras, or of simplicial commutative rings or a similar category with an appropriate variant of a Grothendieck topology. One can also replace presheaves of sets by presheaves of simplicial sets (or of infinity groupoids). Then, in presence of an appropriate homotopic machinery one can develop a notion of derived stack as such a presheaf on the infinity category of derived affine schemes, which is satisfying certain infinite categorical version of a sheaf axiom (and to be algebraic, inductively a sequence of representability conditions). Quillen model categories, Segal categories and quasicategories are some of the most often used tools to formalize this yielding the derived algebraic geometry, introduced by the school of Carlos Simpson, including Andre Hirschowitz, Bertrand Toën, Gabrielle Vezzosi, Michel Vaquié and others; and developed further by Jacob Lurie, Bertrand Toën, and Gabrielle Vezzosi. Another (noncommutative) version of derived algebraic geometry, using A-infinity categories has been developed from early 1990s by Maxim Kontsevich and followers.

Prehistory: Before the 16th Century

Some of the roots of algebraic geometry date back to the work of the Hellenistic Greeks from the 5th century BC. The Delian problem, for instance, was to construct a length x so that the cube of side x contained the same volume as the rectangular box a^2b for given sides a and b. Menaechmus (circa 350 BC) considered the problem geometrically by intersecting the pair of plane conics $ay = x^2$ and $xy = ab$. The later work, in the 3rd century BC, of Archimedes and Apollonius studied more systematically problems on conic sections, and also involved the use of coordinates.The Arab mathematicians were able to solve by purely algebraic means certain cubic equations, and then to interpret the results geometrically. This was done, for instance, by Ibn al-Haytham in the 10th century AD. Subsequently, Persian mathematician Omar Khayyám (born 1048 A.D.) discovered the general method of solving cubic equations by intersecting a parabola with a circle. Each of these early developments in algebraic geometry dealt with questions of finding and describing the intersections of algebraic curves.

Renaissance

Such techniques of applying geometrical constructions to algebraic problems were also adopted by a number of Renaissance mathematicians such as Gerolamo Cardano and Niccolò Fontana "Tartaglia" on their studies of the cubic equation. The geometrical approach to construction problems, rather than the algebraic one, was favored by most 16th and 17th century mathematicians, notably Blaise Pascal who argued against the use of algebraic and analytical methods in geometry. The French mathematicians Franciscus Vieta and later René Descartes and Pierre de Fermat revolutionized the conventional way of thinking about construction problems through the introduction of coordinate geometry. They were interested primarily in the properties of algebraic curves, such as those defined by Diophantine equations (in the case of Fermat), and the algebraic reformulation of the classical Greek works on conics and cubics (in the case of Descartes).

During the same period, Blaise Pascal and Gérard Desargues approached geometry from a different perspective, developing the synthetic notions of projective geometry. Pascal and Desargues also studied curves, but from the purely geometrical point of view: the analog of the Greek ruler and compass construction. Ultimately, the analytic geometry of Descartes and Fermat won out, for it supplied the 18th century mathematicians with concrete quantitative tools needed to study physical problems using the new calculus of

Newton and Leibniz. However, by the end of the 18th century, most of the algebraic character of coordinate geometry was subsumed by the calculus of infinitesimals of Lagrange and Euler.

19th and Early 20th Century

It took the simultaneous 19th century developments of non-Euclidean geometry and Abelian integrals in order to bring the old algebraic ideas back into the geometrical fold. The first of these new developments was seized up by Edmond Laguerre and Arthur Cayley, who attempted to ascertain the generalized metric properties of projective space. Cayley introduced the idea of homogeneous polynomial forms, and more specifically quadratic forms, on projective space. Subsequently, Felix Klein studied projective geometry (along with other types of geometry) from the viewpoint that the geometry on a space is encoded in a certain class of transformations on the space. By the end of the 19th century, projective geometers were studying more general kinds of transformations on figures in projective space. Rather than the projective linear transformations which were normally regarded as giving the fundamental Kleinian geometry on projective space, they concerned themselves also with the higher degree birational transformations. This weaker notion of congruence would later lead members of the 20th century Italian school of algebraic geometry to classify algebraic surfaces up to birational isomorphism.

The second early 19th century development, that of Abelian integrals, would lead Bernhard Riemann to the development of Riemann surfaces. In the same period began the algebraization of the algebraic geometry through commutative algebra. The prominent results in this direction are Hilbert's basis theorem and Hilbert's Nullstellensatz, which are the basis of the connexion between algebraic geometry and commutative algebra, and Macaulay's multivariate resultant, which is the basis of elimination theory. Probably because of the size of the computation which is implied by multivariate resultants, elimination theory was forgotten during the middle of the 20th century until it was renewed by singularity theory and computational algebraic geometry.

20th Century

B. L. van der Waerden, Oscar Zariski and André Weil developed a foundation for algebraic geometry based on contemporary commutative algebra, including valuation theory and the theory of ideals. One of the goals was to give a rigorous framework for proving the results of Italian school of algebraic geometry. In particular, this school used systematically the notion of generic point without any precise definition, which was first given by these authors during the 1930s.

In the 1950s and 1960s Jean-Pierre Serre and Alexander Grothendieck recast the foundations making use of sheaf theory. Later, from about 1960, and largely led by Grothendieck, the idea of schemes was worked out, in conjunction with a very refined apparatus of homological techniques. After a decade of rapid development the field stabilized in the 1970s, and new applications were made, both to number theory and to more classical geometric questions on algebraic varieties, singularities and moduli.

An important class of varieties, not easily understood directly from their defining equations, are the abelian varieties, which are the projective varieties whose points form an abelian group. The prototypical examples are the elliptic curves, which have a rich theory. They were instrumental in the proof of Fermat's last theorem and are also used in elliptic curve cryptography.

In parallel with the abstract trend of the algebraic geometry, which is concerned with general statements about varieties, methods for effective computation with concretely-given varieties have also been developed, which lead to the new area of computational algebraic geometry. One of the founding methods of this area is the theory of Gröbner bases, introduced by Bruno Buchberger in 1965. Another founding method, more specially devoted to real algebraic geometry, is the cylindrical algebraic decomposition, introduced by George E. Collins in 1973.

Analytic Geometry

An **analytic variety** is defined locally as the set of common solutions of several equations involving analytic functions. It is analogous to the included concept of real or complex algebraic variety. Any complex manifold is an analytic variety. Since analytic varieties may have singular points, not all analytic varieties are manifolds.

Modern analytic geometry is essentially equivalent to real and complex algebraic geometry, as has been shown by Jean-Pierre Serre in his paper GAGA, the name of which is French for Algebraic geometry and analytic geometry. Nevertheless, the two fields remain distinct, as the methods of proof are quite different and algebraic geometry includes also geometry in finite characteristic.

GEOMETRIC ALGEBRA

The geometric product was first briefly mentioned by Hermann Grassmann, who was chiefly interested in developing the closely related exterior algebra, which is the geometric algebra of the trivial quadratic form. In 1878, William Kingdon Clifford greatly expanded on Grassmann›s work to form what are now usually called Clifford algebras in his honor (although Clifford himself chose to call them geometric algebras). For several decades, geometric algebras went somewhat ignored, greatly eclipsed by the vector calculus then newly developed to describe electromagnetism. The term geometric algebra was repopularized by Hestenes in the 1960s, who advocated its importance to relativistic physics.

DEFINITION AND NOTATION

Given a finite-dimensional real quadratic space V with a symmetric bilinear form (e.g. the Euclidean or Lorentzian metric) $g : V \times V \to R$, the geometric algebra for this quadratic space is the Clifford algebra $C\ell(V, g)$.

The algebra product is called the geometric product. It is standard to denote the geometric product by juxtaposition (i.e., suppressing any explicit multiplication symbol). The above definition of the geometric algebra is abstract, so we summarize the properties of the geometric product by the following set of axioms. The geometric product has the following properties:

1a=a1, where 1 is the unit and {\displaystyle a} a any element of the algebra (existence of the unit)

a(bc)=(ab)c, where a, b and c are any elements of the algebra (associativity)

a(b+c)=ab+ac and (b+c)a=ba+ca, where a, b and c are any elements of the algebra (distributivity)

$a^2 = g(a,a)1$,where a is any element of the subspace V of the algebra.

Note that in the final property above, the square of a vector need not be nonnegative if g is not positive-definite. An important property of the geometric product is the existence of elements with multiplicative inverse, also known as units. If $a^2 \neq 0$ for some vector a, then a^{-1} exists and is equal to a/a^2. Not every nonzero element of the algebra is necessarily a unit. For example, if u is a vector in V such that $u^2 = 1$, the elements $1 \pm u$ are zero divisors and thus have no inverse: $(1 - u)(1 + u) = 1 - uu = 1 - 1 = 0$. There may also exist nontrivial idempotent elements such as $(1 + u)/2$.

Inner and Exterior Product of Vectors

For vectors a and b, we may write the geometric product of any two vectors a and b as the sum of a symmetric product and an antisymmetric product:

$$ab = \frac{1}{2}(ab + ba) + \frac{1}{2}(ab - ba)$$

Thus we can define the *inner product* of vectors as

$$a \cdot b := g(a, b),$$

so that the symmetric product can be written as

$$\frac{1}{2}(ab + ba) = \frac{1}{2}((a + b)^2 - a^2 - b^2) = a \cdot b,$$

Conversely, g is completely determined by the algebra. The antisymmetric part is the exterior product (also called the outer product) of the two vectors, the product of the contained exterior algebra:

$$a \wedge b := \frac{1}{2}(ab - ba) = -(b \wedge a)$$

The inner and exterior products are associated with familiar concepts from standard vector algebra. Pictorially, a and b are parallel if their geometric product is equal to their inner product, whereas a and b are perpendicular if their geometric product is equal to their exterior product. In a geometric algebra for which the square of any nonzero vector is positive, the inner product of two vectors can be identified with the dot product of standard vector algebra. The exterior product of two vectors can be identified with the signed area enclosed by a parallelogram the sides of which are the vectors. The cross product of two vectors in 3 dimensions with positive-definite quadratic form is closely related to their exterior product.

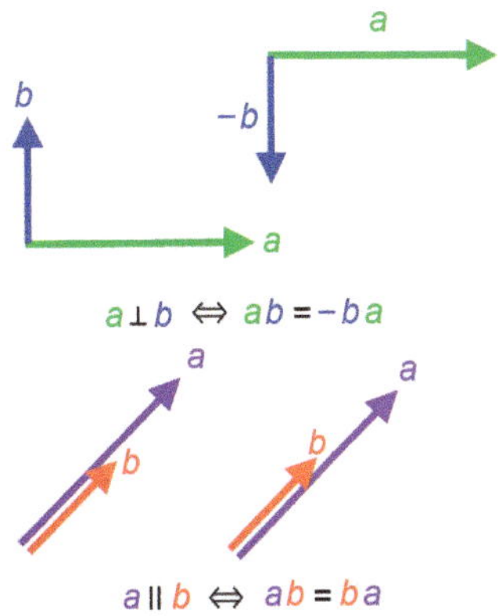

Given two vectors a and b, if the geometric product ab isanticommutative; they are perpendicular (top) because a • b = 0, if it is commutative; they are parallel (bottom) because a ∧ b = 0

Most instances of geometric algebras of interest have a nondegenerate quadratic form. If the quadratic form is fully degenerate, the inner product of any two vectors is always zero, and the geometric algebra is then simply an exterior algebra. Unless otherwise stated, this article will treat only nondegenerate geometric algebras.

The exterior product is naturally extended as a completely antisymmetric, multilinear operator between any number of vectors:

$$a_1 \wedge a_2 \wedge \cdots \wedge a_r = \frac{1}{r!} \sum_{\sigma \in \mathfrak{S}_r} \operatorname{sgn}(\sigma) a_{\sigma(1)} a_{\sigma(2)} \cdots a_{\sigma(r)},$$

where the sum is over all permutations of the indices, with sng(σ) the sign of the permutation.

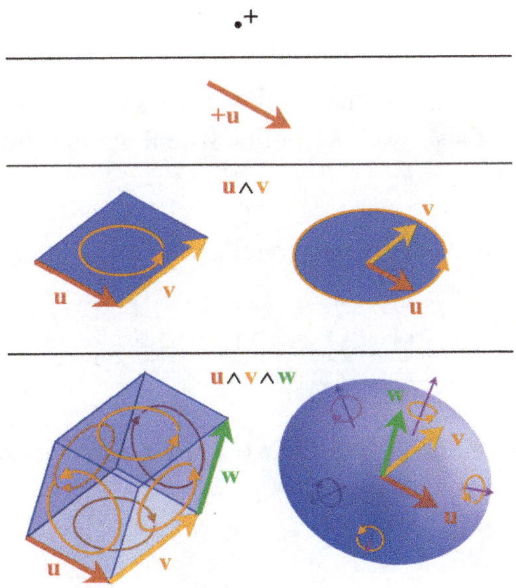

Orientation defined by an ordered set of vectors.

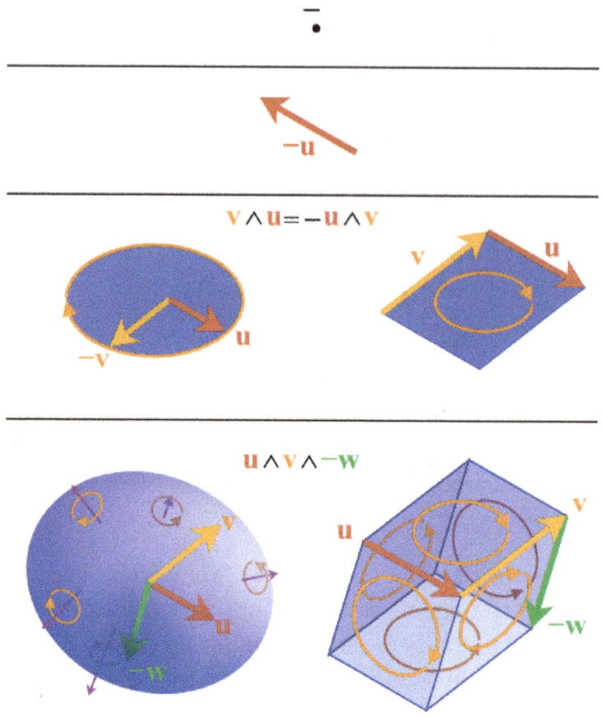

Reversed orientation corresponds to negating the exterior product.

Blades, grades, and canonical basis

A multivector that is the exterior product of r independent vectors ($r \leq n$) is called a blade, and the blade is said to be a multivector of grade r. From the axioms, with closure, every multivector of the geometric algebra is a sum of blades.

Consider a set of r independent vectors $\{a_1, \ldots, a_r\}$ spanning an r-dimensional subspace of the vector space. With these, we can define a real symmetric matrix

$$[\mathbf{A}]_{ij} = a_i \cdot a_j$$

By the spectral theorem, A can be diagonalized to diagonal matrix D by an orthogonal matrix O via

$$\sum_{k,l}[\mathbf{O}]_{ik}[\mathbf{A}]_{kl}[\mathbf{O}^{\mathrm{T}}]_{lj} = \sum_{k,l}[\mathbf{O}]_{ik}[\mathbf{O}]_{jl}[\mathbf{A}]_{kl} = [\mathbf{D}]_{ij}$$

Define a new set of vector s$\{e_1, \ldots, e_r\}$, known as orthogonal basis vectors, to be those transformed by the orthogonal matrix:

$$e_i = \sum_j [\mathbf{O}]_{ij} a_j$$

Since orthogonal transformations preserve inner products, it follows that $e_i \cdot e_j = [\mathbf{D}]_{ij}$ and thus the $\{e_1, \ldots, e_r\}$ are perpendicular. In other words, the geometric product of two distinct vectors $e_i \neq e_j$ is completely specified by their exterior product, or more generally

$$\begin{aligned} e_1 e_2 \cdots e_r &= e_1 \wedge e_2 \wedge \cdots \wedge e_r \\ &= \left(\textstyle\sum_j [\mathbf{O}]_{1j} a_j\right) \wedge \left(\textstyle\sum_j [\mathbf{O}]_{2j} a_j\right) \wedge \cdots \wedge \left(\textstyle\sum_j [\mathbf{O}]_{rj} a_j\right) \\ &= \det[\mathbf{O}] a_1 \wedge a_2 \wedge \cdots \wedge a_r \end{aligned}$$

Therefore, every blade of grade r can be written as a geometric product of r vectors. More generally, if a degenerate geometric algebra is allowed, then the orthogonal matrix is replaced by a block matrix that is orthogonal in the nondegenerate block, and the diagonal matrix has zero-valued entries along the degenerate dimensions. If the new vectors of the nondegenerate subspace are normalized according to

$$\hat{e}_i = \frac{1}{\sqrt{|e_i \cdot e_i|}} e_i,$$

then these normalized vectors must square to +1 or -1. By Sylvester's law of inertia, the total number of +1s and the total number of -1s along the diagonal matrix is invariant. By extension, the total number p of these vectors that square to +1 and the total number q that square to -1 is invariant. (If the degenerate case is allowed, then the total number of basis vectors that square to zero is also invariant.) We denote this algebra g(p,q). For example, g(3,0) models 3- dimensional Euclidean space, g(1,3) relativistic spacetime and g(4,1) a conformal geometric algebra of a 3- dimensional space.

The set of all possible products of n orthogonal basis vectors with indices in increasing order, including 1 as the empty product forms a basis for the entire geometric algebra (an analogue of the PBW theorem). For example, the following is a basis for the geometric algebra g(3,0):

$$\{1, e_1, e_2, e_3, e_1 e_2, e_1 e_3, e_2 e_3, e_1 e_2 e_3\}$$

A basis formed this way is called a canonical basis for the geometric algebra, and any other orthogonal basis for V will produce another canonical basis. Each canonical basis consists of 2^n elements. Every multivector of the geometric algebra can be expressed as a linear combination of the canonical basis elements. If the canonical basis elements are $\{B_i \mid i \in S\}$ with S being an index set, then the geometric product of any two multivectors is

$$\left(\sum_i \alpha_i B_i\right)\left(\sum_j \beta_j B_j\right) = \sum_{i,j} \alpha_i \beta_j B_i B_j.$$

Grade Projection

Using a canonical basis, a graded vector space structure can be established. Elements of the geometric algebra that are simply scalar multiples of 1 are grade-0 blades and are called scalars. Nonzero multivectors that are in the span of $\{e_1,\ldots,e_n\}$ are grade-1 blades and are the ordinary vectors. Multivectors in the span of $\{e_ie_j \mid 1\le i<j\le n\}$ are grade-2 blades and are the bivectors. This terminology continues through to the last grade of n-vectors. Alternatively, grade-n blades are called pseudoscalars, grade-n−1 blades pseudovectors, etc. Many of the elements of the algebra are not graded by this scheme since they are sums of elements of differing grade. Such elements are said to be of mixed grade. The grading of multivectors is independent of the orthogonal basis chosen originally.

A multivector A may be decomposed with the grade-projection operator $\langle A\rangle_r$, which outputs the grade -r portion of A. As a result:

$$A=\sum_{r=0}^{n}\langle A\rangle_r$$

As an example, the geometric product of two vectors $ab=a\cdot b+a\wedge b=\langle ab\rangle_0+\langle ab\rangle_2$ since $\langle ab\rangle_0=a\cdot b$ and $\langle ab\rangle_2=a\wedge b$ and $\langle ab\rangle_i=0$, for i other than 0 and 2.

The decomposition of a multivector A may also be split into those components that are even and those that are odd:

$$A^+=\langle A\rangle_0+\langle A\rangle_2+\langle A\rangle_4+\cdots$$
$$A^-=\langle A\rangle_1+\langle A\rangle_3+\langle A\rangle_5+\cdots$$

This makes the algebra a $\mathbf{Z}_2$-graded algebra or superalgebra with the geometric product. Since the geometric product of two even multivectors is an even multivector, they define an even subalgebra. The even subalgebra of an n-dimensional geometric algebra is isomorphic to a full geometric algebra of (n−1) dimensions. Examples include $\mathcal{G}^+(2,0)\cong\mathcal{G}(0,1)$ and $\mathcal{G}^+(1,3)\cong\mathcal{G}(3,0)$

Representation of Subspaces

Geometric algebra represents subspaces of V as multivectors, and so they coexist in the same algebra with vectors from V. A k-dimensional subspace W of V is represented by taking an orthogonal basis $\{b_1,b_2,\ldots,b_k\}$ and using the geometric product to form the blade $D = b_1b_2\cdots b_k$. There are multiple blades representing W; all those representing W are scalar multiples of D. These blades can be separated into two sets: positive multiples of D and negative multiples of D. The positive multiples of D are said to have the same orientation as D, and the negative multiples the opposite orientation.

Blades are important since geometric operations such as projections, rotations and reflections depend on the factorability via the exterior product that (the restricted class of) n-blades provide but that (the generalized class of) grade-n multivectors do not when $n \ge 4$.

Unit Pseudoscalars

Unit pseudoscalars are blades that play important roles in GA. A unit pseudoscalar for a non-degenerate subspace W of V is a blade that is the product of the members of an orthonormal basis for W. It can be shown that if I and I′ are both unit pseudoscalars for W, then I = ±I′ and $I^2 = \pm 1$.

Suppose the geometric algebra g(n,0) with the familiar positive definite inner product on R^n is formed. Given a plane (2-dimensional subspace) of R^n, one can find an orthonormal basis $\{b_1, b_2\}$ spanning the plane, and thus find a unit pseudoscalar $I = b_1 b_2$ representing this plane. The geometric product of any two vectors in the span of b_1 and b_2 lies in $\{\alpha_0 + \alpha_1 I \mid \alpha_i \in \mathbf{R}\}$, that is, it is the sum of a 0-vector and a 2-vector.

By the properties of the geometric product,

$I^2 = b_1 b_2 b_1 b_2 = -b_1 b_2 b_2 b_1 = -1$. The resemblance to the imaginary unit is not accidental: the subspace $\{\alpha_0 + \alpha_1 I \mid \alpha_i \in \mathbf{R}\}$ is R-algebra isomorphic to the complex numbers. In this way, a copy of the complex numbers is embedded in the geometric algebra for each 2-dimensional subspace of V on which the quadratic form is definite.

It is sometimes possible to identify the presence of an imaginary unit in a physical equation. Such units arise from one of the many quantities in the real algebra that square to −1, and these have geometric significance because of the properties of the algebra and the interaction of its various subspaces.

In $\mathcal{G}(3,0)$, an exceptional case occurs. Given a canonical basis consisting of orthonormal vectors e_i of V, the set of all 2-vectors is generated by $\{e_3 e_2, e_1 e_3, e_2 e_1\}$.

Labelling these i,j and k (momentarily deviating from our uppercase convention), the subspace generated by 0-vectors and 2-vectors is exactly $\{\alpha_0 + i\alpha_1 + j\alpha_2 + k\alpha_3 \mid \alpha_i \in \mathbf{R}\}$. This set is seen to be a subalgebra, and furthermore is R-algebra isomorphic to the quaternions, another important algebraic system.

Dual Basis

Let $\{e_i\}$ be a basis of V i.e. a set of n linearly independent vectors that span the n-dimensional vector space V. The basis that is dual to $\{e_i\}$ is the set of elements of the dual vector space V* that forms a biorthogonal system with this basis, thus being the elements denoted $\{e^i\}$ satisfying

$$e^i \cdot e_j = \delta^i{}_j,$$

where δ is the Kronecker delta.

Given a nondegenerate quadratic form on V, V^* becomes naturally identified with V, and the dual basis may be regarded as elements of V, but are not in general the same set as the original basis.

Given further a GA of V, let

$$\epsilon = e_1 \wedge \cdots \wedge e_n$$

be the pseudoscalar (which does not necessarily square to ± 1) formed from the basis $\{e_i\}$. he dual basis vectors may be constructed as

$$e^i = (-1)^{i-1}(e_1 \wedge \cdots \wedge \check{e}_i \wedge \cdots \wedge e_n)\epsilon^{-1},$$

where the $\check{e}_i$ denotes that the ith basis vector is omitted from the product.

Extensions of the Inner and Exterior Products

It is common practice to extend the exterior product on vectors to the entire algebra. This may be done through the use of the grade projection operator:

$$C \wedge D := \sum_{r,s} \langle \langle C \rangle_r \langle D \rangle_s \rangle_{r+s} \quad \text{(the \textit{exterior product})}$$

This generalization is consistent with the above definition involving antisymmetrization. Another generalization related to the exterior product is the commutator product:

$$C \times D := \tfrac{1}{2}(CD - DC)$$

The regressive product is the dual of the exterior product:

$$C \vee D := ((CI^{-1}) \wedge (DI^{-1}))I$$

with I the unit pseudoscalar of the algebra. The regressive product, like the exterior product, is associative.

The inner product on vectors can also be generalized, but in more than one non-equivalent way. The paper (Dorst 2002) gives a full treatment of several different inner products developed for geometric algebras and their interrelationships, and the notation is taken from there. Many authors use the same symbol as for the inner product of vectors for their chosen extension (e.g. Hestenes and Perwass). No consistent notation has emerged.

Among these several different generalizations of the inner product on vectors are:

$$C \rfloor D := \sum_{r,s} \langle \langle C \rangle_r \langle D \rangle_s \rangle_{s-r} \quad \text{(the \textit{left contraction})}$$

$$C \lfloor D := \sum_{r,s} \langle \langle C \rangle_r \langle D \rangle_s \rangle_{r-s} \quad \text{(the \textit{right contraction})}$$

$$C * D := \sum_{r,s} \langle \langle C \rangle_r \langle D \rangle_s \rangle_0 \quad \text{(the \textit{scalar product})}$$

$$C \bullet D := \sum_{r,s} \langle \langle C \rangle_r \langle D \rangle_s \rangle_{|s-r|} \quad \text{(the "(fat) dot" product)}$$

Dorst (2002) makes an argument for the use of contractions in preference to Hestenes's inner product; they are algebraically more regular and have cleaner geometric interpretations. A number of identities incorporating the contractions are valid without restriction of their inputs. For example,

$$C \rfloor D = (C \wedge (DI^{-1}))I$$
$$C \lfloor D = I((I^{-1}C) \wedge D)$$
$$A \rfloor (B \rfloor C) = (A \wedge B) \rfloor C$$
$$A \bullet B = A \rfloor B + A \lfloor B - A * B.$$

Benefits of using the left contraction as an extension of the inner product on vectors include that the identity $ab = a \cdot b + a \wedge b$ is extended to $aB = a \rfloor B + a \wedge B$ for any vector a and multivector B, and that the projection operation $\mathcal{P}_b(a) = (a \cdot b^{-1})b$ is extended to $\mathcal{P}_B(A) = (A \rfloor B^{-1}) \rfloor B$ for any blade B and any multivector A (with a minor modification to accommodate null B given below).

Terminology Specific to Geometric Algebra

Some terms are used in geometric algebra with a meaning that differs from the use of those terms in other fields of mathematics. Some of these are listed here:

Vector

In GA this refers to an element of the 1-vector subspace unless otherwise clear from the context, despite the entire algebra forming a vector space.

Grade

In GA this refers to a grading as an algebra under the exterior product (an N-grading), and not under the geometric product (which produces a Z_2^n-grading).

Outer product

When used in GA, this is a synonym for the exterior product. It is not the outer product of linear algebra.

Inner product

When used in GA, this generally refers to the symmetric bilinear form on the 1-vector subspace, not the inner product on a normed vector space. Some authors may extend its meaning to the entire algebra, but there is little consensus on this. Even in texts on geometric algebras, the term is not universally used.

Versor

In GA this refers to an element of the algebra that can be constructed as the geometric product of any number of non-null vectors; an element of the Clifford group. Elsewhere, the term may refer to a unit quaternion, analogous to a rotor in GA.

Outermorphism

This term is used only in GA, and refers to a linear map on the vector subspace, extended to apply to the entire algebra by defining it as preserving the exterior product.

GEOMETRIC INTERPRETATION

Projection and Rejection

For any vector a and any invertible vector m,

$$a = amm^{-1} = (a \cdot m + a \wedge m)m^{-1} = a_{\|m} + a_{\perp m},$$

where the **projection** of a onto m (or the parallel part) is

$$a_{\|m} = (a \cdot m)m^{-1}$$

and the **rejection** of a onto m (or the perpendicular part) is

$$a_{\perp m} = a - a_{\|m} = (a \wedge m)m^{-1}.$$

Using the concept of a k-blade B as representing a subspace of V and every multivector ultimately being expressed in terms of vectors, this generalizes to projection of a general multivector onto any invertible k-blade B as

$$\mathcal{P}_B(A) = (A \rfloor B^{-1}) \rfloor B,$$

with the rejection being defined as

$$\mathcal{P}_B^{\perp}(A) = A - \mathcal{P}_B(A).$$

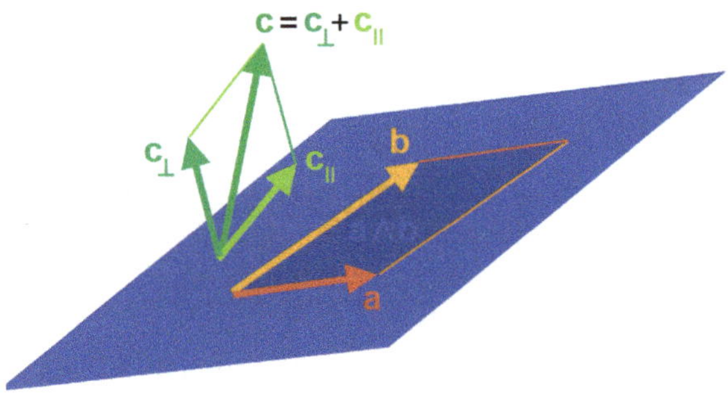

In 3d space, a bivector **a∧b** defines a 2d plane subspace (light blue, extends infinitely in indicated directions). Any vector **c** in the 3-space can be projected onto and rejected normal to the plane, shown respectively by $\mathbf{c}_{\perp}$ and $\mathbf{c}_{\|}$.

The projection and rejection generalize to null blades B by replacing the inverse B^{-1} with the pseudoinverse B^{+} with respect to the contractive product. The outcome of the projection coincides in both cases for non-null blades. For null blades B, the definition of the projection given here with the first contraction rather than the second being onto the pseudoinverse should be used, as only then is the result necessarily in the subspace represented by B. The projection generalizes through linearity to general multivectors A. The projection is not linear in B and does not generalize to objects B that are not blades.

Reflections

The definition of a reflection occurs in two forms in the literature. Several authors work with reflection on a vector (negating all vector components except that parallel to the specifying vector), while others work with reflection along a vector (negating only the component parallel to the specifying vector, or reflection in the hypersurface orthogonal to that vector). Either may be used to build general versor operations, but the former has the advantage that it extends to the algebra in a simpler and algebraically more regular fashion.

Reflection *on* a Vector

The result (c′) of reflecting a vector c on another vector n is to negate the rejection of c. It is akin to reflecting the vector c through the origin, except that the projection of c onto n is not changed. Such an operation is described by

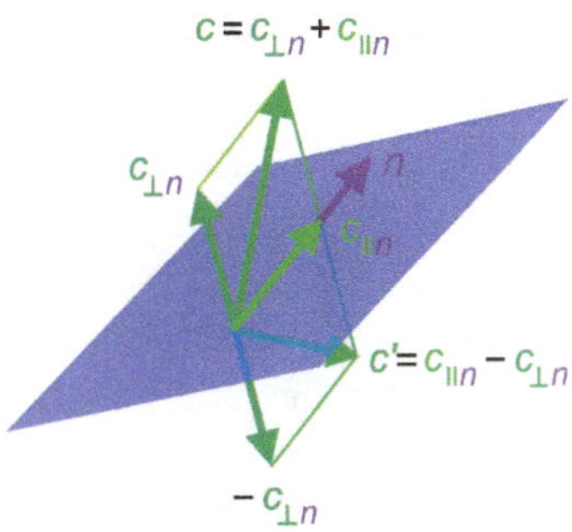

Reflection of vector c on a vector n. The rejection of c on n is negated.

$$c \mapsto c' = ncn^{-1}.$$

Repeating this operation with several different vectors n results in a general versor operation (including both rotations and reflections) of a general multivector A being expressed as

$$A \mapsto NAN^{-1},$$

where N is the geometric product of several vectors. This allows a general definition of any versor N (including both reflections and rotors) as an object that can be expressed as a geometric product of any number of non-null 1-vectors. Such a versor can be applied in a uniform sandwich product as above irrespective of whether it is of even (a proper rotation) or odd grade (an improper rotation i.e. general reflection). The set of all versors with the geometric product as the group operation constitutes the Clifford group of the Clifford algebra $C\ell_{p,q}(\mathbf{R})$.

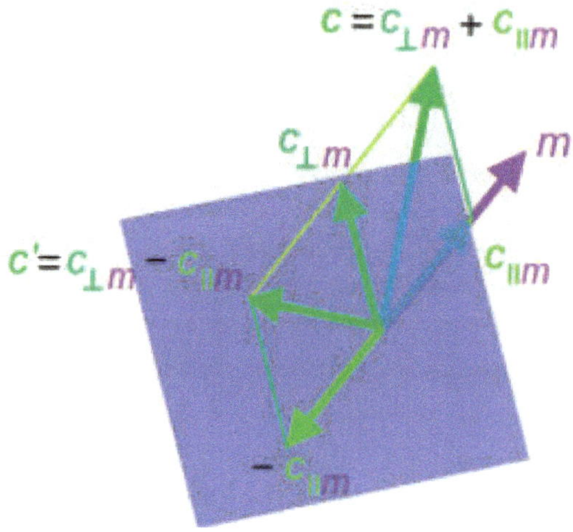

Reflection of vector c along a vector m. Only the component of c parallel to m is negated.

Reflection along a Vector

The reflection (c') of a vector c along a vector m, or equivalently in the hyperplane orthogonal to m, is the same as negating the component of a vector parallel to m. The result of the reflection will be

$$c' = -c_{\|m} + c_{\perp m} = -(c \cdot m)m^{-1} + (c \wedge m)m^{-1} = (-m \cdot c - m \wedge c)m^{-1} = -mcm^{-1}$$

This is not the most general operation that may be regarded as a reflection when the dimension n ≥ 4. A general reflection may be expressed as the composite of any odd number of single-axis reflections. Thus, a general reflection (a') of a vector a may be written

$$a \mapsto a' = -MaM^{-1}$$

where

$$M = pq \cdots r \text{ and} M^{-1} = (pq \cdots r)^{-1} = r^{-1} \cdots q^{-1} p^{-1}.$$

If we define the reflection along a non-null vector m of the product of vectors as the reflection of every vector in the product along the same vector, we get for any product of an odd number of vectors that, by way of example,

$$(abc)' = a'b'c' = (-mam^{-1})(-mbm^{-1})(-mcm^{-1}) = -ma(m^{-1}m)b(m^{-1}m)cm^{-1} = -mabcm^{-1}$$

and for the product of an even number of vectors that

$$(abcd)' = a'b'c'd' = (-mam^{-1})(-mbm^{-1})(-mcm^{-1})(-mdm^{-1}) = mabcdm^{-1}.$$

Using the concept of every multivector ultimately being expressed in terms of vectors, the reflection of a general multivector A using any reflection versor M may be written

$$A \mapsto M\alpha(A)M^{-1},$$

where α is the automorphism of reflection through the origin of the vector space ($v \mapsto -v$) extended through linearity to the whole algebra.

Hypervolume of an n-Parallelotope Spanned by n Vectors

For vectors $\boldsymbol{a}$ and $\boldsymbol{b}$ spanning a parallelogram we have

$$a \wedge b = ((a \wedge b)b^{-1})b = a_{\perp b}b$$

with the result that $\boldsymbol{a} \wedge \boldsymbol{b}$ is linear in the product of the "altitude" and the "base" of the parallelogram, that is, its area.

Similar interpretations are true for any number of vectors spanning an n-dimensional parallelotope; the exterior product of vectors a_1, a_2, ..., a_n, that is $\bigwedge_{i=1}^{n} a_i$, has a magnitude equal to the volume of the n-parallelotope. An n-vector doesn't necessarily have a shape of a parallelotope – this is a convenient visualization. It could be any shape, although the volume equals that of the parallelotope.

Rotations

If we have a product of vectors $\boldsymbol{R} = \boldsymbol{a}_1\boldsymbol{a}_2 \cdots \boldsymbol{a}_r$ then we denote the reverse as

$$\boldsymbol{R}^\dagger = (\boldsymbol{a}_1\boldsymbol{a}_2 \cdots \boldsymbol{a}_r)^\dagger = \boldsymbol{a}_r \cdots \boldsymbol{a}_2\boldsymbol{a}_1.$$

As an example, assume that $\boldsymbol{R} = \boldsymbol{ab}$ we get

$$\boldsymbol{RR}^\dagger = \boldsymbol{abba} = \boldsymbol{ab}^2\boldsymbol{a} = \boldsymbol{a}^2\boldsymbol{b}^2 = \boldsymbol{ba}^2\boldsymbol{b} = \boldsymbol{baab} = \boldsymbol{R}^\dagger\boldsymbol{R}.$$

Scaling R so that $RR^\dagger = 1$ then

$$(\boldsymbol{RvR}^\dagger)^2 = \boldsymbol{Rv}^2\boldsymbol{R}^\dagger = \boldsymbol{v}^2\boldsymbol{RR}^\dagger = \boldsymbol{v}^2$$

so $\boldsymbol{RvR}^\dagger$ leaves the length of $\boldsymbol{v}$ unchanged. We can also show that

$$(\boldsymbol{Rv}_1\boldsymbol{R}^\dagger) \cdot (\boldsymbol{Rv}_2\boldsymbol{R}^\dagger) = \boldsymbol{v}_1 \cdot \boldsymbol{v}_2$$

so the transformation $RvR^\dagger$ preserves both length and angle. It therefore can be identified as a rotation or rotoreflection; R is called a rotor if it is a proper rotation (as it is if it can be expressed as a product of an even number of vectors) and is an instance of what is known in GA as a versor (presumably for historical reasons).

There is a general method for rotating a vector involving the formation of a multivector of the form $\boldsymbol{R} = e^{-\frac{\boldsymbol{B}\theta}{2}}$ that produces a rotation θ in the plane and with the orientation defined by a 2-blade $\boldsymbol{B}$.

Rotors are a generalization of quaternions to n-D spaces.

For more about reflections, rotations and "sandwiching" products like $RvR^\dagger$ see Plane of rotation.

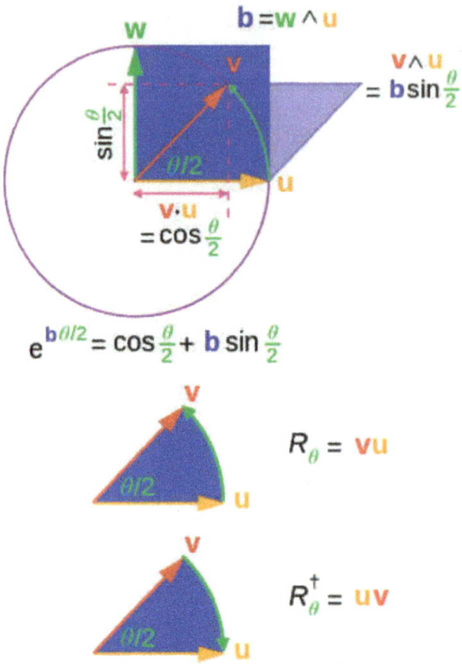

A rotor that rotates vectors in a plane rotates vectors through angle θ, that is $x \mapsto R_\theta x R_\theta^\dagger$ is a rotation of x through angle θ. The angle between u and v is θ/2. Similar interpretations are valid for a general multivector X instead of the vector x.

Linear Functions

An important class of functions of multivectors are the linear functions mapping multivectors to multivectors. The geometric algebra of an n-dimensional vector space is spanned by a basis of 2^n elements. If a multivector is represented by a $2^n \times 1$ real column matrix in terms of a basis, then all linear transformations of the multivector can be expressed as the matrix multiplication by a $2^n \times 2^n$ real matrix. However, such a general linear transformation allows arbitrary exchanges among grades, such as a "rotation" of a scalar into a vector, which has no clear geometric interpretation.

A general linear transformation from vectors to vectors is of interest. Thought there is no unique generalization of these transformations to the entire geometric algebra without further restriction, a natural choice is that of the outermorphism of the linear transformation. It extends the transformation through the expression of a multivector as a sum-of-products-of-vectors within the algebra and applies a given linear map to each vector in the expression. If f is a linear function that maps vectors to vectors, then its outermorphism is the function that obeys the rule

$$\underline{f}(a_1 \wedge a_2 \wedge \cdots \wedge a_r) = f(a_1) \wedge f(a_2) \wedge \cdots \wedge f(a_r)$$

for a blade, extended to the whole algebra through linearity.

In the cases of reflections and rotations, their outermorphisms have a particularly simple algebraic form. Specifically, a mapping of vectors of the form

$$V \to V : a \mapsto RaR^{-1},$$

extends to the outermorphism

$$\mathcal{G}(V) \to \mathcal{G}(V) : A \mapsto RAR^{-1}.$$

EXAMPLES AND APPLICATIONS

Intersection of a Line and a Plane

We may define the line parametrically by $p = t + \alpha v$ where p and t are position vectors for points T and P and v is the direction vector for the line.

Then

$B \wedge (p - q) = 0$ and $B \wedge (t + \alpha v - q) = 0$

so

$$\alpha = \frac{B \wedge (q - t)}{B \wedge v}$$

and

$$p = t + \left(\frac{B \wedge (q - t)}{B \wedge v}\right) v.$$

Rotating Systems

The mathematical description of rotational forces such as torque and angular momentum often makes use of the cross product of vector calculus in three dimensions with a convention of orientation (handedness).

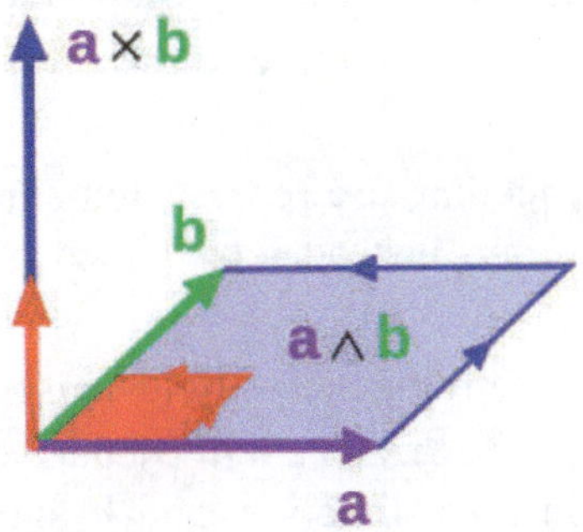

The cross product in relation to the exterior product. In red are the unit normal vector, and the "parallel" unit bivector.

The cross product can be viewed in terms of the exterior product allowing a more natural geometric interpretation of the cross product as a bivector using the dual relationship

$$a \times b = -I(a \wedge b).$$

For example, torque is generally defined as the magnitude of the perpendicular force component times distance, or work per unit angle.

Suppose a circular path in an arbitrary plane containing orthonormal vectors $\hat{u}$ and $\hat{v}$ parameterized by angle.

$$\mathbf{r} = r(\hat{u}\cos\theta + \hat{v}\sin\theta) = r\hat{u}(\cos\theta + \hat{u}\hat{v}\sin\theta)$$

By designating the unit bivector of this plane as the imaginary number

$$i = \hat{u}\hat{v} = \hat{u} \wedge \hat{v}$$
$$i^2 = -1$$

this path vector can be conveniently written in complex exponential form

$$\mathbf{r} = r\hat{u}e^{i\theta}$$

and the derivative with respect to angle is

$$\frac{d\mathbf{r}}{d\theta} = r\hat{u}ie^{i\theta} = \mathbf{r}i$$

So the torque, the rate of change of work W, due to a force F, is

$$\tau = \frac{dW}{d\theta} = F \cdot \frac{dr}{d\theta} = F \cdot (\mathbf{r}i)$$

Unlike the cross product description of torque, $\tau = \mathbf{r} \times F$, the geometric algebra description does not introduce a vector in the normal direction; a vector that does not exist in two and that is not unique in greater than three dimensions. The unit bivector describes the plane and the orientation of the rotation, and the sense of the rotation is relative to the angle between the vectors $\hat{u}$ and $\hat{v}$.

Electrodynamics and Special Relativity

In physics, the main applications are the geometric algebra of Minkowski 3+1 spacetime, $C\ell_{1,3}$, called spacetime algebra (STA). or less commonly, $C\ell_3$, called the algebra of physical space (APS) where $C\ell_3$ is isomorphic to the even subalgebra of the 3+1 Clifford algebra, $C\ell0_3$,1.

While in STA points of spacetime are represented simply by vectors, in APS, points of (3+1)-dimensional spacetime are instead represented by paravectors: a 3-dimensional vector (space) plus a 1-dimensional scalar (time).

In spacetime algebra the electromagnetic field tensor has a bivector representation $F = (E + icB)\gamma_0$. Here, the $i = \gamma_0\gamma_1\gamma_2\gamma_3$ is the unit pseudoscalar (four-dimensional) volume element, γ_0 is the unit vector in time direction, and E and B are the classic electric and magnetic field vectors (with a zero time component). Using the four-current J, Maxwell's equations then become

Formulation	Homogeneous equations	Non-homogeneous equations
Fields	$DF = \mu_0 J$	
	$D \wedge F = 0$	$D \cdot F = \mu_0 J$
Potentials (any gauge)	$F = D \wedge A$	$D \cdot D \wedge A = \mu_0 J$
Potentials (Lorenz gauge)	$F = DA$ $D \cdot A = 0$	$D^2 A = \mu_0 J$

In geometric calculus, juxtapositioning of vectors such as in DF indicate the geometric product and can be decomposed into parts as $DF = D \cdot F + D \wedge F$. Here D is the covector derivative in any spacetime and reduces to ∇ in flat spacetime. Where $\bigtriangledown$ plays a role in Minkowski 4- spacetime which is synonymous to the role of ∇ in Euclidean 3- space and is related to the D'Alembertian by $\Box = \bigtriangledown^2$. Indeed, given an observer represented by a future pointing timelike vector γ0 we have

$$\gamma_0 \cdot \bigtriangledown = \frac{1}{c}\frac{\partial}{\partial t}$$

$$\gamma_0 \wedge \bigtriangledown = \nabla$$

Boosts in this Lorentzian metric space have the same expression e^{β} as rotation in Educlidean space, where β is the bivector generated by the time and the space directions involved, whereas in the Euclidean case it is the bivector generated by the two space directions, strengthening the "analogy" to almost identity.

Relationship with Other Formalisms

$\mathcal{G}(3,0)$ may be directly compared to vector algebra. The even subalgebra of $\mathcal{G}(2,0)$ is isomorphic to the complex numbers, as may be seen by writing a vector P in terms of its components in an orthonormal basis and left multiplying by the basis vector e_1 yielding

$$Z = e_1 P = e_1(xe_1 + ye_2) = x(1) + y(e_1 e_2),$$

where we identify $i \mapsto e_1 e_2$ since

$$(e_1 e_2)^2 = e_1 e_2 e_1 e_2 = -e_1 e_1 e_2 e_2 = -1.$$

Similarly, the even subalgebra of $\mathcal{G}(3,0)$ with basis $\{1, e_2 e_3, e_3 e_1, e_1 e_2\}$ is isomorphic to the quaternions as may be seen by identifying $i \mapsto -e_2 e_3$, $j \mapsto -e_3 e_1$ and $k \mapsto -e_1 e_2$.

Every associative algebra has a matrix representation; the Pauli matrices are a representation of $\mathcal{G}(3,0)$ and the Dirac matrices are a representation of $\mathcal{G}(1,3)$ showing the equivalence with matrix representations used by physicists.

Geometric Calculus

Geometric calculus extends the formalism to include differentiation and integration including differential geometry and differential forms.

Essentially, the vector derivative is defined so that the GA version of Green's theorem is true,

$$\int_A dA\, \nabla f = \oint_{\partial A} dx\, f$$

and then one can write

$$\nabla f = \nabla \cdot f + \nabla \wedge f$$

as a geometric product, effectively generalizing Stokes' theorem (including the differential form version of it).

In $1D$ when A is a curve with endpoints a and b, then

$$\int_A dA\, \nabla f = \oint_{\partial A} dx\, f$$

reduces to

$$\int_a^b dx\, \nabla f = \int_a^b dx \cdot \nabla f = \int_a^b df = f(b) - f(a)$$

or the fundamental theorem of integral calculus.

Also developed are the concept of vector manifold and geometric integration theory (which generalizes Cartan's differential forms).

Conformal Geometric Algebra (CGA)

A compact description of the current state of the art is provided by Bayro-Corrochano and Scheuermann (2010), which also includes further references, in particular to Dorst et al (2007). Other useful references are Li (2008) and Bayro (2010).

Working within GA, Euclidean space $\mathcal{E}^3$ is embedded projectively in the CGA $\mathcal{G}^{4,1}$ via the identification of Euclidean points with 1D subspaces in the 4D null cone of the 5D CGA vector subspace, and adding a point at infinity. This allows all conformal transformations to be done as rotations and reflections and is covariant, extending incidence relations of projective geometry to circles and spheres.

Specifically, we add orthogonal basis vectors e_+ and e_- such that $e_+{}^2 = +1$ and $e_-{}^2 = -1$ to the basis of $\mathcal{G}(3,0)$ and identify null vectors

$n_\infty = e_- + e_+$ as an ideal point (point at infinity) (see Compactification) and

$n_o = \frac{1}{2}(e_- - e_+)$ as the point at the origin, giving

$n_\infty \cdot n_o = -1$.

This procedure has some similarities to the procedure for working with homogeneous coordinates in projective geometry and in this case allows the modeling of Euclidean transformations as orthogonal transformations.

A fast changing and fluid area of GA, CGA is also being investigated for applications to relativistic physics.

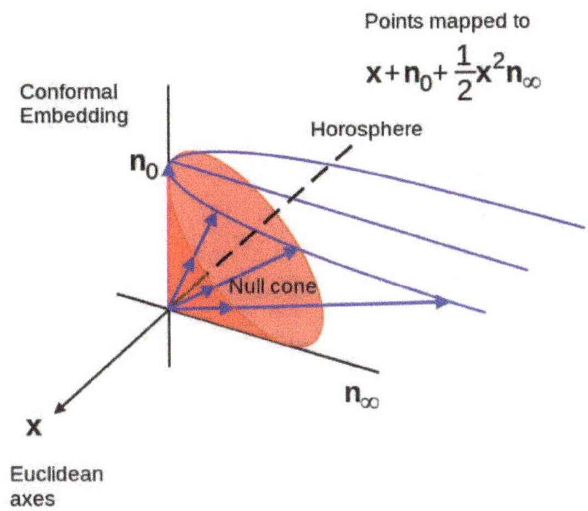

Before the 20th century

Although the connection of geometry with algebra dates as far back at least to Euclid's Elements in the 3rd century B.C. (see Greek geometric algebra), GA in the sense used in this article was not developed until 1844, when it was used in a systematic way to describe the geometrical properties and transformations of a space. In that year, Hermann Grassmann introduced the idea of a geometrical algebra in full generality as a certain calculus (analogous to the propositional calculus) that encoded all of the geometrical information of a space. Grassmann's algebraic system could be applied to a number of different kinds of spaces, the chief among them being Euclidean space, affine space, and projective space. Following Grassmann, in 1878 William Kingdon Clifford examined Grassmann's algebraic system alongside the quaternions of William Rowan Hamilton in (Clifford 1878). From his point of view, the quaternions described certain transformations (which he called rotors), whereas Grassmann's algebra described certain properties (or Strecken such as length, area, and volume). His contribution was to define a new product — the **geometric product** — on an existing Grassmann algebra, which realized the quaternions as living within that algebra. Subsequently Rudolf Lipschitz in 1886 generalized Clifford's interpretation of the quaternions and applied them to the geometry of rotations in n dimensions. Later these developments would lead other 20th-century mathematicians to formalize and explore the properties of the Clifford algebra.

Nevertheless, another revolutionary development of the 19th-century would completely overshadow the geometric algebras: that of vector analysis, developed independently by Josiah Willard Gibbs and Oliver Heaviside. Vector analysis was motivated by James Clerk Maxwell›s studies of electromagnetism, and specifically the need to express and manipulate conveniently certain differential equations. Vector analysis had a certain intuitive appeal compared to the rigors of the new algebras. Physicists and mathematicians alike readily adopted it as their geometrical toolkit of choice, particularly following the influential 1901 textbook Vector Analysis by Edwin Bidwell Wilson, following lectures of Gibbs.

In more detail, there have been three approaches to geometric algebra: quaternionic analysis, initiated by Hamilton in 1843 and geometrized as rotors by Clifford in 1878; geometric algebra, initiated by Grassmann in 1844; and vector analysis, developed out of quaternionic analysis in the late 19th century by Gibbs and Heaviside. The legacy of quaternionic analysis in vector analysis can be seen in the use of i, j, k to indicate the basis vectors of $\mathbf{R}^3$: it is being thought of as the purely imaginary quaternions. From the perspective of geometric algebra, quaternions can be identified as $C\ell^0_{3,0}(\mathbf{R})$, the even part of the Clifford algebra on Euclidean 3-space, which unifies the three approaches.

20th century and present

Progress on the study of Clifford algebras quietly advanced through the twentieth century, although largely due to the work of abstract algebraists such as Hermann Weyl and Claude Chevalley. The geometrical approach to geometric algebras has seen a number of 20th-century revivals. In mathematics, Emil Artin's Geometric Algebra discusses the algebra associated with each of a number of geometries, including affine geometry, projective geometry, symplectic geometry, and orthogonal geometry. In physics, geometric algebras have been revived as a new way to do classical mechanics and electromagnetism, together with more advanced topics such as quantum mechanics and gauge theory. David Hestenes reinterpreted the Pauli and Dirac matrices as vectors in ordinary space and spacetime, respectively, and has been a primary contemporary advocate for the use of geometric algebra.

In computer graphics and robotics, geometric algebras have been revived in order to efficiently represent rotations and other transformations. For applications of GA in robotics (screw theory, kinematics and dynamics using versors), computer vision, control and neural computing (geometric learning) see Bayro (2010).

Software

GA is a very application-oriented subject. There is a reasonably steep initial learning curve associated with it, but this can be eased somewhat by the use of applicable software.

The following is a list of freely available software that does not require ownership of commercial software or purchase of any commercial products for this purpose:

- GA Viewer Fontijne, Dorst, Bouma & Mann
- galgebra modules for sympy Alan Bromborsky

The link provides a manual, introduction to GA and sample material as well as the software.

- CLUViz Perwass

Software allowing script creation and including sample visualizations, manual and GA introduction.

- Gaigen Fontijne

For programmers, this is a code generator with support for C, C++, C# and Java.

- Cinderella Visualizations Hitzer and Dorst.
- Gaalop Standalone GUI-Application that uses the Open-Source Computer Algebra Software Maxima to break down CLUViz code into C/C++ or Java code.
- Gaalop Precompiler Precompiler based on Gaalop integrated with CMake.
- Gaalet, C++ Expression Template Library Seybold.
- Versor Colapinto, A lightweight templated C++ Library with an OpenGL interface for efficient geometric algebra programming in arbitrary metrics, including conformal
- Clifford Algebra with Mathematica clifford.m

NON-EUCLIDEAN GEOMETRY

In mathematics, **non-Euclidean geometry** consists of two geometries based on axioms closely related to those specifying Euclidean geometry. As Euclidean geometry lies at the intersection of metric geometry and affine geometry, non-Euclidean geometry arises when either the metric requirement is relaxed, or the parallel postulate is replaced with an alternative one. In the latter case one obtains hyperbolic geometry and elliptic geometry, the traditional non-Euclidean geometries. When the metric requirement is relaxed, then there are affine planes associated with the planar algebras which give rise to kinematic geometries that have also been called non-Euclidean geometry.

The essential difference between the metric geometries is the nature of parallel lines. Euclid›s fifth postulate, the parallel postulate, is equivalent to Playfair›s postulate, which states that, within a two-dimensional plane, for any given line ℓ and a point A, which is not on ℓ, there is exactly one line through A that does not intersect ℓ. In hyperbolic geometry, by contrast, there are infinitely many lines through A not intersecting ℓ, while in elliptic geometry, any line through A intersects ℓ.

Another way to describe the differences between these geometries is to consider two straight lines indefinitely extended in a two-dimensional plane that are both perpendicular to a third line:

- In Euclidean geometry the lines remain at a constant distance from each other (meaning that a line drawn perpendicular to one line at any point will intersect the other line and the length of the line segment joining the points of intersection remains constant) and are known as parallels.
- In hyperbolic geometry they "curve away" from each other, increasing in distance as one moves further from the points of intersection with the common perpendicular; these lines are often called ultraparallels.
- In elliptic geometry the lines "curve toward" each other and intersect.

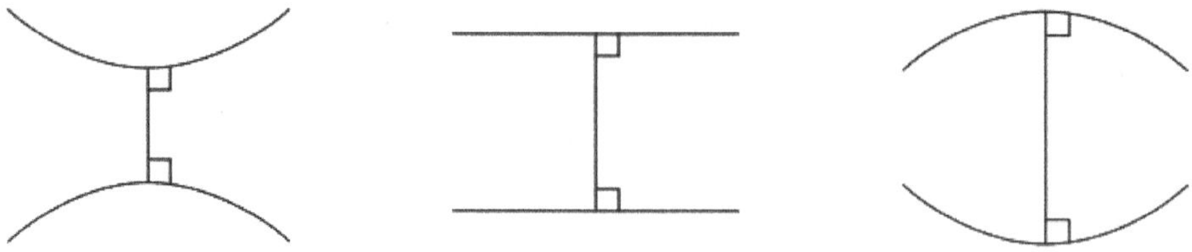

Behavior of lines with a common perpendicular in each of the three types of geometry

Early History

While Euclidean geometry, named after the Greek mathematician Euclid, includes some of the oldest known mathematics, non-Euclidean geometries were not widely accepted as legitimate until the 19th century.

The debate that eventually led to the discovery of the non-Euclidean geometries began almost as soon as Euclid's work Elements was written. In the Elements, Euclid began with a limited number of assumptions (23 definitions, five common notions, and five postulates) and sought to prove all the other results (propositions) in the work. The most notorious of the postulates is often referred to as "Euclid's Fifth Postulate," or simply the "parallel postulate", which in Euclid's original formulation is:

If a straight line falls on two straight lines in such a manner that the interior angles on the same side are together less than two right angles, then the straight lines, if produced indefinitely, meet on that side on which are the angles less than the two right angles.

Other mathematicians have devised simpler forms of this property. Regardless of the form of the postulate, however, it consistently appears to be more complicated than Euclid's other postulates:

1. To draw a straight line from any point to any point.
2. To produce extend a finite straight line continuously in a straight line.
3. To describe a circle with any centre and distance radius.
4. That all right angles are equal to one another.

For at least a thousand years, geometers were troubled by the disparate complexity of the fifth postulate, and believed it could be proved as a theorem from the other four. Many attempted to find a proof by contradiction, including Ibn al-Haytham (Alhazen, 11th century), Omar Khayyám (12th century), Nasīr al-Dīn al-Tūsī (13th century), and Giovanni Girolamo Saccheri (18th century).

The theorems of Ibn al-Haytham, Khayyam and al-Tusi on quadrilaterals, including the Lambert quadrilateral and Saccheri quadrilateral, were "the first few theorems of the hyperbolic and the elliptic geometries." These theorems along with their alternative postulates, such as Playfair's axiom, played an important role in the later development of non-Euclidean geometry. These early attempts at challenging the fifth postulate had a considerable influence on its development among later European geometers, including

Witelo, Levi ben Gerson, Alfonso, John Wallis and Saccheri. All of these early attempts made at trying to formulate non-Euclidean geometry however provided flawed proofs of the parallel postulate, containing assumptions that were essentially equivalent to the parallel postulate. These early attempts did, however, provide some early properties of the hyperbolic and elliptic geometries.

Khayyam, for example, tried to derive it from an equivalent postulate he formulated from "the principles of the Philosopher" (Aristotle): "Two convergent straight lines intersect and it is impossible for two convergent straight lines to diverge in the direction in which they converge."Khayyam then considered the three cases right, obtuse, and acute that the summit angles of a Saccheri quadrilateral can take and after proving a number of theorems about them, he correctly refuted the obtuse and acute cases based on his postulate and hence derived the classic postulate of Euclid which he didn't realize was equivalent to his own postulate. Another example is al-Tusi's son, Sadr al-Din (sometimes known as "Pseudo-Tusi"), who wrote a book on the subject in 1298, based on al-Tusi's later thoughts, which presented another hypothesis equivalent to the parallel postulate. "He essentially revised both the Euclidean system of axioms and postulates and the proofs of many propositions from the Elements."His work was published in Rome in 1594 and was studied by European geometers, including Saccheri who criticised this work as well as that of Wallis.

Giordano Vitale, in his book Euclide restituo (1680, 1686), used the Saccheri quadrilateral to prove that if three points are equidistant on the base AB and the summit CD, then AB and CD are everywhere equidistant.

In a work titled Euclides ab Omni Naevo Vindicatus (Euclid Freed from All Flaws), published in 1733, Saccheri quickly discarded elliptic geometry as a possibility (some others of Euclid's axioms must be modified for elliptic geometry to work) and set to work proving a great number of results in hyperbolic geometry.

He finally reached a point where he believed that his results demonstrated the impossibility of hyperbolic geometry. His claim seems to have been based on Euclidean presuppositions, because no logical contradiction was present. In this attempt to prove Euclidean geometry he instead unintentionally discovered a new viable geometry, but did not realize it.

In 1766 Johann Lambert wrote, but did not publish, Theorie der Parallellinien in which he attempted, as Saccheri did, to prove the fifth postulate. He worked with a figure that today we call a Lambert quadrilateral, a quadrilateral with three right angles (can be considered half of a Saccheri quadrilateral). He quickly eliminated the possibility that the fourth angle is obtuse, as had Saccheri and Khayyam, and then proceeded to prove many theorems under the assumption of an acute angle. Unlike Saccheri, he never felt that he had reached a contradiction with this assumption. He had proved the non-Euclidean result that the sum of the angles in a triangle increases as the area of the triangle decreases, and this led him to speculate on the possibility of a model of the acute case on a sphere of imaginary radius. He did not carry this idea any further.

At this time it was widely believed that the universe worked according to the principles of Euclidean geometry.

Discovery of Non-Euclidean Geometry

The beginning of the 19th century would finally witness decisive steps in the creation of non-Euclidean geometry. Circa 1813, Carl Friedrich Gauss and independently around 1818, the German professor of law Ferdinand Karl Schweikart had the germinal ideas of non-Euclidean geometry worked out, but neither published any results. Then, around 1830, the Hungarian mathematician János Bolyai and the Russian mathematician Nikolai Ivanovich Lobachevsky separately published treatises on hyperbolic geometry. Consequently, hyperbolic geometry is called Bolyai-Lobachevskian geometry, as both mathematicians, independent of each other, are the basic authors of non-Euclidean geometry. Gauss mentioned to Bolyai's father, when shown the younger Bolyai's work, that he had developed such a geometry several years before, though he did not publish. While Lobachevsky created a non-Euclidean geometry by negating the parallel postulate, Bolyai worked out a geometry where both the Euclidean and the hyperbolic geometry are possible depending on a parameter k. Bolyai ends his work by mentioning that it is not possible to decide through mathematical reasoning alone if the geometry of the physical universe is Euclidean or non-Euclidean; this is a task for the physical sciences.

Bernhard Riemann, in a famous lecture in 1854, founded the field of Riemannian geometry, discussing in particular the ideas now called manifolds, Riemannian metric, and curvature. He constructed an infinite family of geometries which are not Euclidean by giving a formula for a family of Riemannian metrics on the unit ball in Euclidean space. The simplest of these is called elliptic geometry and it is considered to be a non-Euclidean geometry due to its lack of parallel lines.

By formulating the geometry in terms of a curvature tensor, Riemann allowed non-Euclidean geometry to be applied to higher dimensions.

Terminology

It was Gauss who coined the term "non-Euclidean geometry". He was referring to his own work which today we call hyperbolic geometry. Several modern authors still consider "non-Euclidean geometry" and "hyperbolic geometry" to be synonyms.

Arthur Cayley noted that distance between points inside a conic could be defined in terms of logarithm and the projective cross-ratio function. The method has become called the Cayley-Klein metric because Felix Klein exploited it to describe the non-euclidean geometries in articles in 1871 and 73 and later in book form. The Cayley-Klein metrics provided working models of hyperbolic and elliptic metric geometries, as well as Euclidean geometry.

Klein is responsible for the terms "hyperbolic" and "elliptic" (in his system he called Euclidean geometry "parabolic", a term which generally fell out of use). His influence has led to the current usage of the term "non-Euclidean geometry" to mean either "hyperbolic" or "elliptic" geometry.

There are some mathematicians who would extend the list of geometries that should be called "non-Euclidean" in various ways.

Axiomatic Basis of Non-Euclidean Geometry

Euclidean geometry can be axiomatically described in several ways. Unfortunately, Euclid's original system of five postulates (axioms) is not one of these as his proofs relied on several unstated assumptions which should also have been taken as axioms. Hilbert's system consisting of 20 axioms most closely follows the approach of Euclid and provides the justification for all of Euclid's proofs. Other systems, using different sets of undefined terms obtain the same geometry by different paths. In all approaches, however, there is an axiom which is logically equivalent to Euclid's fifth postulate, the parallel postulate. Hilbert uses the Playfair axiom form, while Birkhoff, for instance, uses the axiom which says that "there exists a pair of similar but not congruent triangles." In any of these systems, removal of the one axiom which is equivalent to the parallel postulate, in whatever form it takes, and leaving all the other axioms intact, produces absolute geometry. As the first 28 propositions of Euclid (in The Elements) do not require the use of the parallel postulate or anything equivalent to it, they are all true statements in absolute geometry.

To obtain a non-Euclidean geometry, the parallel postulate (or its equivalent) must be replaced by its negation. Negating the Playfair's axiom form, since it is a compound statement (... there exists one and only one ...), can be done in two ways:

- Either there will exist more than one line through the point parallel to the given line or there will exist no lines through the point parallel to the given line. In the first case, replacing the parallel postulate (or its equivalent) with the statement In a plane, given a point P and a line ℓ not passing through P, there exist two lines through P which do not meet ℓ" and keeping all the other axioms, yields hyperbolic geometry.
- The second case is not dealt with as easily. Simply replacing the parallel postulate with the statement, "In a plane, given a point P and a line ℓ not passing through P, all the lines through P meet ℓ", does not give a consistent set of axioms. This follows since parallel lines exist in absolute geometry, but this statement says that there are no parallel lines. This problem was known (in a different guise) to Khayyam, Saccheri and Lambert and was the basis for their rejecting what was known as the "obtuse angle case". In order to obtain a consistent set of axioms which includes this axiom about having no parallel lines, some of the other axioms must be tweaked. The adjustments to be made depend upon the axiom system being used. Among others these tweaks will have the effect of modifying Euclid's second postulate from the statement that line segments can be extended indefinitely to the statement that lines are unbounded. Riemann's

elliptic geometry emerges as the most natural geometry satisfying this axiom.

Models of Non-Euclidean Geometry

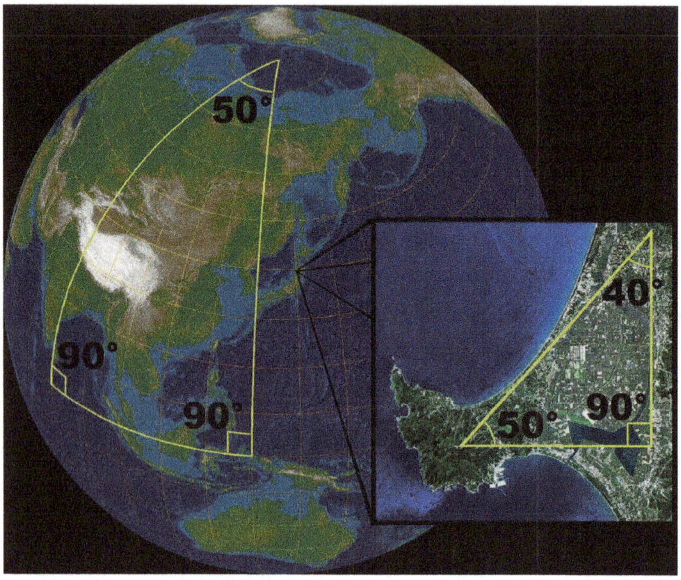

On a sphere, the sum of the angles of a triangle is not equal to 180°. The surface of a sphere is not a Euclidean space, but locally the laws of the Euclidean geometry are good approximations. In a small triangle on the face of the earth, the sum of the angles is very nearly 180°.

Two dimensional Euclidean geometry is modelled by our notion of a "flat plane."

Elliptic Geometry

The simplest model for elliptic geometry is a sphere, where lines are "great circles" (such as the equator or the meridians on a globe), and points opposite each other (called antipodal points) are identified (considered to be the same). This is also one of the standard models of the real projective plane. The difference is that as a model of elliptic geometry a metric is introduced permitting the measurement of lengths and angles, while as a model of the projective plane there is no such metric.

In the elliptic model, for any given line ℓ and a point A, which is not on ℓ, all lines through A will intersect ℓ.

Hyperbolic Geometry

Even after the work of Lobachevsky, Gauss, and Bolyai, the question remained: "Does such a model exist for hyperbolic geometry?". The model for hyperbolic geometry was answered by Eugenio Beltrami, in 1868, who first showed that a surface called the pseudosphere has the appropriate curvature to model a portion of hyperbolic space and in a second paper in the same year, defined the Klein model which models the entirety of hyperbolic space, and used this to show that Euclidean geometry and hyperbolic geometry were equiconsistent so that hyperbolic geometry was logically consistent if and only if Euclidean geometry was. (The reverse implication follows from the horosphere model of Euclidean geometry.)

In the hyperbolic model, within a two-dimensional plane, for any given line ℓ and a point A, which is not on ℓ, there are infinitely many lines through A that do not intersect ℓ.

In these models the concepts of non-Euclidean geometries are being represented by Euclidean objects in a Euclidean setting. This introduces a perceptual distortion wherein the straight lines of the non-Euclidean geometry are being represented by Euclidean curves which visually bend. This "bending" is not a property of the non-Euclidean lines, only an artifice of the way they are being represented.

Three-Dimensional Non-Euclidean Geometry

In three dimensions, there are eight models of geometries. There are Euclidean, elliptic, and hyperbolic geometries, as in the two-dimensional case; mixed geometries that are partially Euclidean and partially hyperbolic or spherical; twisted versions of the mixed geometries; and one unusual geometry that is completely anisotropic (i.e. every direction behaves differently).

Uncommon Properties

Euclidean and non-Euclidean geometries naturally have many similar properties, namely those which do not depend upon the nature of parallelism. This commonality is the subject of absolute geometry (also called neutral geometry). However, the properties which distinguish one geometry from the others are the ones which have historically received the most attention.

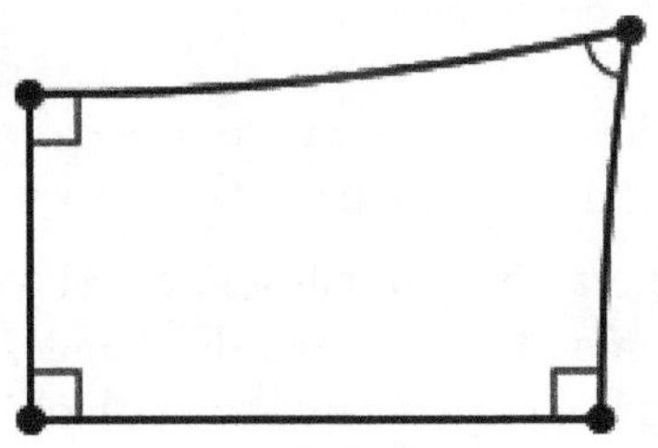

Lambert quadrilateral in hyperbolic geometry

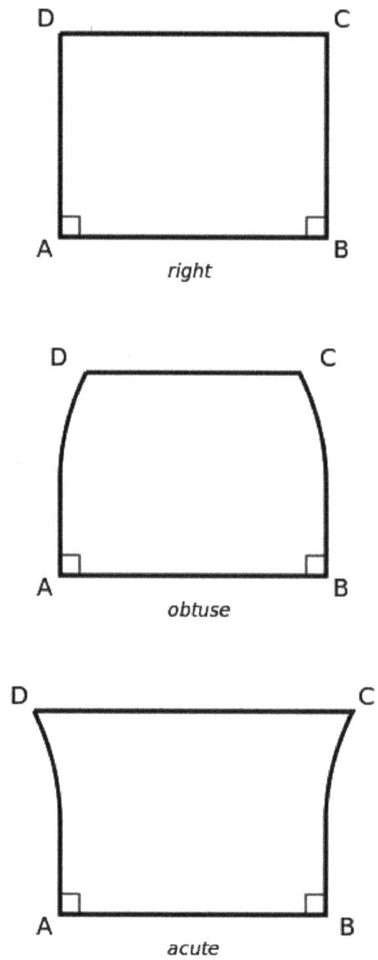

Saccheri quadrilaterals in the three geometries

Besides the behavior of lines with respect to a common perpendicular, mentioned in the introduction, we also have the following:

- A Lambert quadrilateral is a quadrilateral which has three right angles. The fourth angle of a Lambert quadrilateral is acute if the geometry is hyperbolic, a right angle if the geometry is Euclidean or obtuse if the geometry is elliptic. Consequently, rectangles exist (a statement equivalent to the parallel postulate) only in Euclidean geometry.
- A Saccheri quadrilateral is a quadrilateral which has two sides of equal length, both perpendicular to a side called the base. The other two angles of a Saccheri quadrilateral are called the summit angles and they have equal measure. The summit angles of a Saccheri quadrilateral are acute if the geometry is hyperbolic, right angles if the geometry is Euclidean and obtuse angles if the geometry is elliptic.

- The sum of the measures of the angles of any triangle is less than 180° if the geometry is hyperbolic, equal to 180° if the geometry is Euclidean, and greater than 180° if the geometry is elliptic. The defect of a triangle is the numerical value (180° - sum of the measures of the angles of the triangle). This result may also be stated as: the defect of triangles in hyperbolic geometry is positive, the defect of triangles in Euclidean geometry is zero, and the defect of triangles in elliptic geometry is negative.

Importance

Before the models of a non-Euclidean plane were presented by Beltrami, Klein, and Poincaré, Euclidean geometry stood unchallenged as the mathematical model of space. Furthermore, since the substance of the subject in synthetic geometry was a chief exhibit of rationality, the Euclidean point of view represented absolute authority.

The discovery of the non-Euclidean geometries had a ripple effect which went far beyond the boundaries of mathematics and science. The philosopher Immanuel Kant's treatment of human knowledge had a special role for geometry. It was his prime example of synthetic a priori knowledge; not derived from the senses nor deduced through logic — our knowledge of space was a truth that we were born with. Unfortunately for Kant, his concept of this unalterably true geometry was Euclidean. Theology was also affected by the change from absolute truth to relative truth in the way that mathematics is related to the world around it, that was a result of this paradigm shift.

Non-Euclidean geometry is an example of a scientific revolution in the history of science, in which mathematicians and scientists changed the way they viewed their subjects. Some geometers called Lobachevsky the "Copernicus of Geometry" due to the revolutionary character of his work.

The existence of non-Euclidean geometries impacted the intellectual life of Victorian England in many ways and in particular was one of the leading factors that caused a re-examination of the teaching of geometry based on Euclid's Elements. This curriculum issue was hotly debated at the time and was even the subject of a book, Euclid and his Modern Rivals, written by Charles Lutwidge Dodgson (1832–1898) better known as Lewis Carroll, the author of Alice in Wonderland.

Planar Algebras

In analytic geometry a plane is described with Cartesian coordinates : $C = \{ (x,y) : x, y \in \mathbb{R} \}$. The points are sometimes identified with complex numbers $z = x + y\,\varepsilon$ where $\varepsilon^2 \in \{-1, 0, 1\}$.

The Euclidean plane corresponds to the case $\varepsilon^2 = -1$ since the modulus of z is given by

$$zz^* = (x + y\epsilon)(x - y\epsilon) = x^2 + y^2$$

and this quantity is the square of the Euclidean distance between z and the origin. For instance, $\{z \mid z\, z^* = 1\}$ is the unit circle.

For planar algebra, non-Euclidean geometry arises in the other cases. When $\varepsilon^2 = +1$, then z is a split-complex number and conventionally **j**replaces epsilon. Then

$$zz^* = (x + y\mathbf{j})(x - y\mathbf{j}) = x^2 - y^2$$

and {z | z z* = 1} is the unit hyperbola.

When $\varepsilon^2 = 0$, then z is a dual number.

This approach to non-Euclidean geometry explains the non-Euclidean angles: the parameters of slope in the dual number plane and hyperbolic angle in the split-complex plane correspond to angle in Euclidean geometry. Indeed, they each arise in polar decomposition of a complex number z.

Kinematic Geometries

Hyperbolic geometry found an application in kinematics with the physical cosmology introduced by Hermann Minkowski in 1908. Minkowski introduced terms like worldline and proper time into mathematical physics. He realized that the submanifold, of events one moment of proper time into the future, could be considered a hyperbolic space of three dimensions. Already in the 1890s Alexander Macfarlane was charting this submanifold through his Algebra of Physics and hyperbolic quaternions, though Macfarlane did not use cosmological language as Minkowski did in 1908. The relevant structure is now called the hyperboloid model of hyperbolic geometry.

The non-Euclidean planar algebras support kinematic geometries in the plane. For instance, the split-complex number $z = e^{aj}$ can represent a spacetime event one moment into the future of a frame of reference of rapidity a. Furthermore, multiplication by z amounts to a Lorentz boost mapping the frame with rapidity zero to that with rapidity a.

Kinematic study makes use of the dual numbers $z = x + y\epsilon, \quad \epsilon^2 = 0,$ to represent the classical description of motion in absolute time and space: The equations $x' = x + vt, \quad t' = t$ are equivalent to a shear mapping in linear algebra:

$$\begin{pmatrix} x' \\ t' \end{pmatrix} = \begin{pmatrix} 1 & v \\ 0 & 1 \end{pmatrix} \begin{pmatrix} x \\ t \end{pmatrix}.$$

With dual numbers the mapping is $t' + x'\epsilon = (1 + v\epsilon)(t + x\epsilon) = t + (x + vt)\epsilon.$

Another view of special relativity as a non-Euclidean geometry was advanced by E. B. Wilson and Gilbert Lewis in Proceedings of the American Academy of Arts and Sciences in 1912. They revamped the analytic geometry implicit in the split-complex number algebra into synthetic geometry of premises and deductions.

Fiction

Non-Euclidean geometry often makes appearances in works of science fiction and fantasy.

- In 1895 H. G. Wells published the short story "The Remarkable Case of

Davidson's Eyes". To appreciate this story one should know how antipodal points on a sphere are identified in a model of the elliptic plane. In the story, in the midst of a thunderstorm, Sidney Davidson sees "Waves and a remarkably neat schooner" while working in an electrical laboratory at Harlow Technical College. At the story's close Davidson proves to have witnessed H.M.S. Fulmar off Antipodes Island.

- Non-Euclidean geometry is sometimes connected with the influence of the 20th century horror fiction writer H. P. Lovecraft. In his works, many unnatural things follow their own unique laws of geometry: In Lovecraft›s Cthulhu Mythos, the sunken city of R›lyeh is characterized by its non-Euclidean geometry. It is heavily implied this is achieved as a side effect of not following the natural laws of this universe rather than simply using an alternate geometric model, as the sheer innate wrongness of it is said to be capable of driving those who look upon it insane.
- The main character in Robert Pirsig›s Zen and the Art of Motorcycle Maintenance mentioned Riemannian Geometry on multiple occasions.
- In The Brothers Karamazov, Dostoevsky discusses non-Euclidean geometry through his main character Ivan.
- Christopher Priest's novel Inverted World describes the struggle of living on a planet with the form of a rotating pseudosphere.
- Robert Heinlein›s The Number of the Beast utilizes non-Euclidean geometry to explain instantaneous transport through space and time and between parallel and fictional universes.
- Alexander Bruce›s Antichamber uses non-Euclidean geometry to create a minimal, Escher-like world, where geometry and space follow unfamiliar rules.
- Zeno Rogue›s HyperRogue is a roguelike game set on the hyperbolic plane, allowing the player to experience many properties of this geometry. Many mechanics, quests, and locations are strongly dependent on the features of hyperbolic geometry.
- In the Renegade Legion science fiction setting for FASA›s wargame, role-playing-game and fiction, faster-than-light travel and communications is possible through the use of Hsieh Ho›s Polydimensional Non-Euclidean Geometry, published sometime in the middle of the 22nd century.
- In Ian Stewart›s Flatterland the protagonist Victoria Line visit all kinds of non-Euclidean worlds.
- In Jean-Pierre Petit›s Here's looking at Euclid (and not looking at Euclid) Archibald Higgins stumbles upon spherical geometry

RIEMANNIAN GEOMETRY

Riemannian geometry is the branch of differential geometry that studies Riemannian manifolds, smooth manifolds with a Riemannian metric, i.e. with an inner product on the tangent space at each point that varies smoothly from point to point. This gives, in particular, local notions of angle, length of curves, surface area and volume. From those, some other global quantities can be derived by integrating local contributions.

Riemannian geometry originated with the vision of Bernhard Riemann expressed in his inaugural lecture Ueber die Hypothesen, welche der Geometrie zu Grunde liegen (On the Hypotheses on which Geometry is based). It is a very broad and abstract generalization of the differential geometry of surfaces in $\mathbf{R}^3$. Development of Riemannian geometry resulted in synthesis of diverse results concerning the geometry of surfaces and the behavior of geodesics on them, with techniques that can be applied to the study of differentiable manifolds of higher dimensions. It enabled the formulation of Einstein's general theory of relativity, made profound impact on group theory and representation theory, as well as analysis, and spurred the development of algebraic and differential topology.

Bernhard Riemann

Riemannian geometry was first put forward in generality by Bernhard Riemann in the 19th century. It deals with a broad range of geometries whose metric properties vary from point to point, including the standard types of Non-Euclidean geometry.

Any smooth manifold admits a Riemannian metric, which often helps to solve problems of differential topology. It also serves as an entry level for the more complicated structure of pseudo-Riemannian manifolds, which (in four dimensions) are the main objects of the theory of general relativity. Other generalizations of Riemannian geometry include

Finsler geometry.

There exists a close analogy of differential geometry with the mathematical structure of defects in regular crystals. Dislocations and Disclinations produce torsions and curvature.

The following articles provide some useful introductory material:

- Metric tensor
- Riemannian manifold
- Levi-Civita connection
- Curvature
- Curvature tensor
- List of differential geometry topics
- Glossary of Riemannian and metric geometry

Classical Theorems in Riemannian Geometry

What follows is an incomplete list of the most classical theorems in Riemannian geometry. The choice is made depending on its importance, beauty, and simplicity of formulation. Most of the results can be found in the classic monograph by Jeff Cheeger and D. Ebin (see below).

The formulations given are far from being very exact or the most general. This list is oriented to those who already know the basic definitions and want to know what these definitions are about.

General Theorems

- **Gauss–Bonnet theorem** The integral of the Gauss curvature on a compact 2-dimensional Riemannian manifold is equal to $2\pi\chi(M)$ where $\chi(M)$ denotes the Euler characteristic of M. This theorem has a generalization to any compact even-dimensional Riemannian manifold, see generalized Gauss-Bonnet theorem.
- **Nash embedding theorems** also called fundamental theorems of Riemannian geometry. They state that every Riemannian manifold can be isometrically embedded in a Euclidean space $\mathbf{R}^n$.

Geometry in Large

In all of the following theorems we assume some local behavior of the space (usually formulated using curvature assumption) to derive some information about the global structure of the space, including either some information on the topological type of the manifold or on the behavior of points at “sufficiently large” distances.

Pinched Sectional Curvature

- **Sphere theorem.** If M is a simply connected compact n-dimensional Riemannian manifold with sectional curvature strictly pinched between 1/4 and 1 then M is diffeomorphic to a sphere.

- **Cheeger›s finiteness theorem.** Given constants C, D and V, there are only finitely many (up to diffeomorphism) compact n-dimensional Riemannian manifolds with sectional curvature $|K| \leq C$, diameter $\leq D$ and volume $\geq V$.
- **Gromov's almost flat manifolds.** There is an $\varepsilon_n > 0$ such that if an n-dimensional Riemannian manifold has a metric with sectional curvature $|K| \leq \varepsilon_n$ and diameter ≤ 1 then its finite cover is diffeomorphic to a nil manifold.

Sectional Curvature Bounded Below

- **Cheeger-Gromoll›s Soul theorem.** If M is a non-compact complete non-negatively curved n-dimensional Riemannian manifold, then M contains a compact, totally geodesic submanifold S such that M is diffeomorphic to the normal bundle of S (S is called the **soul** of M.) In particular, if M has strictly positive curvature everywhere, then it is diffeomorphic to $\mathbf{R}^n$. G. Perelman in 1994 gave an astonishingly elegant/short proof of the Soul Conjecture: M is diffeomorphic to $\mathbf{R}^n$ if it has positive curvature at only one point.
- **Gromov›s Betti number theorem.** There is a constant C = C(n) such that if M is a compact connected n-dimensional Riemannian manifold with positive sectional curvature then the sum of its Betti numbers is at most C.
- **Grove–Petersen's finiteness theorem.** Given constants C, D and V, there are only finitely many homotopy types of compact n-dimensional Riemannian manifolds with sectional curvature $K \geq C$, diameter $\leq D$ and volume $\geq V$.

Sectional Curvature Bounded Above

- The **Cartan–Hadamard theorem** states that a complete simply connected Riemannian manifold M with nonpositive sectional curvature is diffeomorphic to the Euclidean space $\mathbf{R}^n$ with n = dim M via the exponential map at any point. It implies that any two points of a simply connected complete Riemannian manifold with nonpositive sectional curvature are joined by a unique geodesic.
- The geodesic flow of any compact Riemannian manifold with negative sectional curvature is ergodic.
- If M is a complete Riemannian manifold with sectional curvature bounded above by a strictly negative constant k then it is a CAT(k) space. Consequently, its fundamental group $\Gamma = \pi_1(M)$ is Gromov hyperbolic. This has many implications for the structure of the fundamental group:
    - it is finitely presented;
    - the word problem for Γ has a positive solution;
    - the group Γ has finite virtual cohomological dimension;
    - it contains only finitely many conjugacy classes of elements of finite order;

- the abelian subgroups of Γ are virtually cyclic, so that it does not contain a subgroup isomorphic to **Z**×**Z**.

Ricci Curvature Bounded Below

- **Myers theorem.** If a compact Riemannian manifold has positive Ricci curvature then its fundamental group is finite.
- **Bochner›s formula.** If a compact Riemannian n-manifold has non-negative Ricci curvature, then its first Betti number is at most n, with equality if and only if the Riemannian manifold is a flat torus.
- **Splitting theorem.** If a complete n-dimensional Riemannian manifold has nonnegative Ricci curvature and a straight line (i.e. a geodesic that minimizes distance on each interval) then it is isometric to a direct product of the real line and a complete (n-1)-dimensional Riemannian manifold that has nonnegative Ricci curvature.
- **Bishop–Gromov inequality.** The volume of a metric ball of radius r in a complete n-dimensional Riemannian manifold with positive Ricci curvature has volume at most that of the volume of a ball of the same radius r in Euclidean space.
- **Gromov›s compactness theorem.** The set of all Riemannian manifolds with positive Ricci curvature and diameter at most D is pre-compact in the Gromov-Hausdorff metric.

Negative Ricci Curvature

- The isometry group of a compact Riemannian manifold with negative Ricci curvature is discrete.
- Any smooth manifold of dimension n ≥ 3 admits a Riemannian metric with negative Ricci curvature. (This is not true for surfaces.)

Positive Scalar Curvature

- The n-dimensional torus does not admit a metric with positive scalar curvature.
- If the injectivity radius of a compact n-dimensional Riemannian manifold is ≥ π then the average scalar curvature is at most n(n-1).

REFERENCES

1. Baylis, W. E., ed. (1996), Clifford (Geometric) Algebra with Applications to Physics, Mathematics, and Engineering, Birkhäuser
2. Baylis, W. E. (2002), Electrodynamics: A Modern Geometric Approach (2 ed.), Birkhäuser, ISBN 978-0-8176-4025-5
3. Bourbaki, Nicolas (1980), Eléments de Mathématique. Algèbre, Ch. 9 "Algèbres de Clifford": Hermann

4. Dorst, Leo (2002), The inner products of geometric algebra, Boston, MA: Birkhäuser Boston, p. 35–46
5. Hestenes, David (1999), New Foundations for Classical Mechanics (2 ed.), Springer Verlag, ISBN 978-0-7923-5302-7
6. Hestenes, David (1966). Space-time Algebra. New York: Gordon and Breach. ISBN 978-0-677-01390-9. OCLC 996371.
7. Lasenby, J.; Lasenby, A. N.; Doran, C. J. L. (2000), "A Unified Mathematical Language for Physics and Engineering in the 21st Century" (PDF), Philosophical Transactions of the Royal Society of London, pp. 1–18, 358 (A 358): 21, Bibcode:2000RSPTA.358...21L, doi:10.1098/rsta.2000.0517
8. Doran, Chris; Lasenby, Anthony (2003). Geometric algebra for physicists (PDF). Cambridge University Press. ISBN 978-0-521-71595-9.
9. Macdonald, Alan (2011). Linear and Geometric Algebra. Charleston: CreateSpace. ISBN 9781453854938. OCLC 704377582.
10. J Bain (2006). "Spacetime structuralism: §5 Manifolds vs. geometric algebra". In Dennis Dieks. The ontology of spacetime. Elsevier. p. 54 ff. ISBN 978-0-444-52768-4.
11. Bayro-Corrochano, Eduardo (2010). Geometric Computing for Wavelet Transforms, Robot Vision, Learning, Control and Action. Springer Verlag.
12. Vince, John A. (2008), Geometric Algebra for Computer Graphics, Springer, ISBN 978-1-84628-996-5
13. Christian Perwass (2009). Geometric Algebra with Applications in Engineering. Springer Science & Business Media. ISBN 978-3-540-89068-3.
14. Ron Goldman (2010). Rethinking Quaternions: Theory and Computation. Morgan & Claypool Publishers. Part III. Rethinking Quaternions and Clifford Algebras. ISBN 978-1-60845-420-4.
15. John Snygg (2011). A New Approach to Differential Geometry using Clifford's Geometric Algebra. Springer Science & Business Media. ISBN 978-0-8176-8282-8.
16. Dietmar Hildenbrand (2013). Foundations of Geometric Algebra Computing. Springer Science & Business Media. ISBN 978-3-642-31793-4.
17. F. Klein, Über die sogenannte nichteuklidische Geometrie, Mathematische Annalen, 4(1871).
18. The Euclidean plane is still referred to as "parabolic" in the context of conformal geometry: see Uniformization theorem.
19. for instance, Manning 1963 and Yaglom 1968
20. a 21st axiom appeared in the French translation of Hilbert's Grundlagen der Geometrie according to Smart 1997, pg. 416
21. (Smart 1997, pg.366)
22. while only two lines are postulated, it is easily shown that there must be an infinite number of such lines.

23. Book I Proposition 27 of Euclid's Elements
24. William Thurston. Three-dimensional geometry and topology. Vol. 1. Edited by Silvio Levy. Princeton Mathematical Series, 35. Princeton University Press, Princeton, NJ, 1997. x+311 pp. ISBN 0-691-08304-5 (in depth explanation of the eight geometries and the proof that there are only eight)
25. Imre Toth, "Gott und Geometrie: Eine viktorianische Kontroverse," Evolutionstheorie und ihre Evolution, Dieter Henrich, ed. (Schriftenreihe der Universität Regensburg, band 7, 1982) pp. 141–204.
26. Trudeau 1987, p. vii-viii
27. Bell, E. T. (1986). Men of Mathematics. Touchstone Books. p. 294. ISBN 978-0-671-62818-5. Author attributes this quote to another mathematician, William Kingdon Clifford.
28. This is a quote from G. B. Halsted's translator's preface to his 1914 translation of The Theory of Parallels: "What Vesalius was to Galen, what Copernicus was to Ptolemy that was Lobachevsky to Euclid." — W. K. Clifford
29. (Richards 1988)
30. Isaak Yaglom (1968) Complex Numbers in Geometry, translated by E. Primrose from 1963 Russian original, appendix "Non-Euclidean geometries in the plane and complex numbers", pp 195–219, Academic Press, N.Y.
31. Richard C. Tolman (2004) Theory of Relativity of Motion, page 194, §180 Non-Euclidean angle, §181 Kinematical interpretation of angle in terms of velocity
32. Hermann Minkowski (1908–9). "Space and Time" (Wikisource).
33. Scott Walter (1999) Non-Euclidean Style of Special Relativity
34. Isaak Yaglom (1979) A simple non-Euclidean geometry and its physical basis : an elementary account of Galilean geometry and the Galilean principle of relativity, Springer ISBN 0-387-90332-1
35. Edwin B. Wilson & Gilbert N. Lewis (1912) "The Space-time Manifold of Relativity. The Non-Euclidean Geometry of Mechanics and Electromagnetics" Proceedings of the American Academy of Arts and Sciences 48:387–507
36. Synthetic Spacetime, a digest of the axioms used, and theorems proved, by Wilson and Lewis. Archived by WebCite
37. "The Call of Cthulhu".
38. "HyperRogue website".
39. Jean-Pierre Petit. "Here's Looking at Euclid". savoir-sans-frontieres. Retrieved 30 August 2015.
40. Dieudonné, Jean (1972). "The historical development of algebraic geometry". The American Mathematical Monthly. 79 (8): 827–866. doi:10.2307/2317664. JSTOR 2317664.
41. Kline, M. (1972) Mathematical Thought from Ancient to Modern Times (Volume 1). Oxford University Press. pp. 108, 90.

42. Kline, M. (1972) Mathematical Thought from Ancient to Modern Times (Volume 1). Oxford University Press. p. 193.
43. Kline, M. (1972) Mathematical Thought from Ancient to Modern Times (Volume 1). Oxford University Press. pp. 193–195.
44. Kline, M. (1972) Mathematical Thought from Ancient to Modern Times (Volume 1). Oxford University Press. p. 279.
45. A witness of this oblivion is the fact that Van der Waerden removed the chapter on elimination theory from the third edition (and all the subsequent ones) of his treatise Moderne algebra (in German).
46. Drton, Mathias; Sturmfels, Bernd; Sullivant, Seth (2009). Lectures on Algebraic Statistics. Springer. ISBN 978-3-7643-8904-8.
47. Falb, Peter (1990). Methods of Algebraic Geometry in Control Theory Part II Multivariable Linear Systems and Projective Algebraic Geometry. Springer. ISBN 978-0-8176-4113-9.
48. Allen Tannenbaum (1982), Invariance and Systems Theory: Algebraic and Geometric Aspects, Lecture Notes in Mathematics, volume 845, Springer-Verlag, ISBN 9783540105657
49. Selig, J.M. (2005). Geometric Fundamentals of Robotics. Springer. ISBN 978-0-387-20874-9.
50. Tsfasman, Michael A.; Vlădu, Serge G.; Nogin, Dmitry (1990). Algebraic Geometric Codes Basic Notions. American Mathematical Soc. ISBN 978-0-8218-7520-9.
51. Barry A. Cipra (2007), Algebraic Geometers See Ideal Approach to Biology, SIAM News, Volume 40, Number 6
52. Jüttler, Bert; Piene, Ragni (2007). Geometric Modeling and Algebraic Geometry. Springer. ISBN 978-3-540-72185-7.
53. Cox, David A.; Katz, Sheldon (1999). Mirror Symmetry and Algebraic Geometry. American Mathematical Soc. ISBN 978-0-8218-2127-5.
54. Blume, L. E.; Zame, W. R. (1994). "The algebraic geometry of perfect and sequential equilibrium" (PDF). Econometrica. 62 (4): 783–794. JSTOR 2951732.
55. Kenyon, Richard; Okounkov, Andrei; Sheffield, Scott (2003). "Dimers and Amoebae". arXiv:math-ph/0311005 .
56. Fordy, Allan P. (1990). Soliton Theory A Survey of Results. Manchester University Press. ISBN 978-0-7190-1491-8.
57. Cox, David A.; Sturmfels, Bernd. Manocha, Dinesh N., ed. Applications of Computational Algebraic Geometry. American Mathematical Soc. ISBN 978-0-8218-6758-7.

CHAPTER 4

Contemporary Branches of Geometry

Chapter Contents

There are Euclidean (or plane, or flat) geometry and Non-Euclidean (or curved) geometries, and among the latter, there's the separate study of negative curvature and positive curvature. The notions of curvature are advanced, but among these, you most often hear people refer to hyperbolic geometry and spherical geometry as examples. It's hard to say that these are branches of geometry, however: They're just terms which lump together qualitatively similar notions based on which type of space you want your geometry to live on.

INTRODUCTION

Absolute geometry is a geometry based on an axiom system for Euclidean geometry with the parallel postulate removed and none of its alternatives used in place of it. The term was introduced by János Bolyai in 1832. It is sometimes referred to as neutral geometry, as it is neutral with respect to the parallel postulate.

PROPERTIES

It might be imagined that absolute geometry is a rather weak system, but that is not the case. Indeed, in Euclid's *Elements*, the first 28 Propositions and Proposition 31 avoid using the parallel postulate, and therefore are valid in absolute geometry. One can also prove in absolute geometry the exterior angle theorem (an exterior angle of a triangle is larger than either of the remote angles), as well as the Saccheri–Legendre theorem, which states that the sum of the measures of the angles in a triangle has at most 180°.

Proposition 31 is the construction of a parallel line to a given line through a point not on the given line. As the proof only requires the use of Proposition 27 (the Alternate Interior Angle Theorem), it is a valid construction in absolute geometry. More precisely, given any line l and any point P not on l, there is *at least* one line through P which is parallel to l. This can be proved using a familiar construction: given a line l and a point P not on l, drop the perpendicular m from P to l, then erect a perpendicular n to m through P. By the alternate interior angle theorem, l is parallel to n. (The alternate interior angle theorem states that if lines a and b are cut by a transversal t such that there is a pair of congruent alternate interior angles, then a and b are parallel.) The foregoing construction, and the alternate interior angle theorem, do not depend on the parallel postulate and are therefore valid in absolute geometry.

In absolute geometry it is also provable that two lines perpendicular to the same line cannot intersect (which makes the two lines parallel by definition of parallel lines), proving that the summit angles of an Saccheri quadrilateral cannot be obtuse , and that spherical geometry is not an absolute geometry.

Relation to other Geometries

The theorems of absolute geometry hold in hyperbolic geometry, which is a non-Euclidean geometry, as well as in Euclidean geometry.

Absolute geometry is inconsistent with elliptic geometry: in that theory, there are no parallel lines at all, but it is a theorem of absolute geometry that parallel lines do exist.

Absolute geometry is an extension of ordered geometry, and thus, all theorems in ordered geometry hold in absolute geometry. The converse is not true. Absolute geometry assumes the first four of Euclid's Axioms (or their equivalents), to be contrasted with affine geometry, which does not assume Euclid's third and fourth axioms. (3: "To describe a circle with any centre and distance radius.", 4: "That all right angles are equal to one another.") Ordered geometry is a common foundation of both absolute and affine geometry.

The geometry of special relativity has been developed starting with nine axioms and eleven propositions of absolute geometry. The authors Edwin B. Wilson and Gilbert N. Lewis then proceed beyond absolute geometry when they introduce hyperbolic rotation as the transformation relating two frames of reference.

Incompleteness

Absolute geometry is an incomplete axiomatic system, in the sense that one can add extra independent axioms without making the axiom system inconsistent. One can extend absolute geometry by adding different axioms about parallel lines and get incompatible but consistent axiom systems, giving rise to Euclidean or hyperbolic geometry. Thus every theorem of absolute geometry is a theorem of hyperbolic geometry and Euclidean geometry. However the converse is not true.

AFFINE GEOMETRY

In mathematics, **affine geometry** is what remains of Euclidean geometry, when not using (mathematicians often say "when forgetting") the metric notions of distance and angle. As the notion of *parallel lines* is one of the main properties that is independent of any metric, affine geometry is often considered as the study of parallel lines. Therefore, Playfair's axiom (*given a line L and a point P not on L, there is exactly one line parallel to L that passes through P*) is fundamental in affine geometry. Comparisons of figures in affine geometry are made with affine transformations, which are mappings that preserve alignment of points and parallelism of lines.

Affine geometry can be developed in two ways that are essentially equivalent.

In synthetic geometry, an affine space is a set of *points* to which is associated a set of lines, which satisfy some axioms (such as Playfair's axiom).

Affine geometry can also be developed on the basis of linear algebra. In this context an affine space is a set of *points* equipped with a set of *transformations* (that is bijective mappings), the translations, which forms a vector space (over a given field, commonly the real numbers), and such that for any given ordered pair of points there is a unique translation sending the first point to the second; the composition of two translations is their sum in the vector space of the translations.

In more concrete terms, this amounts to having an operation that associates to any ordered pair of points a vector and another operation that allows translation of a point by a vector to give another point; these operations are required to satisfy a number of axioms (notably that two successive translations have the effect of translation by the sum vector). By choosing any point as origin, the points are in one-to-one correspondence with the vectors, but there is no preferred choice for the origin; thus an affine space may be viewed as obtained from its associated vector space by forgetting the origin (zero vector).

Although this article only discusses affine spaces, the notion of forgetting the metric is much more general, and can be applied to arbitrary manifolds, in general. This extension of the notion of affine spaces to manifolds in general is developed in the article on the affine connection.

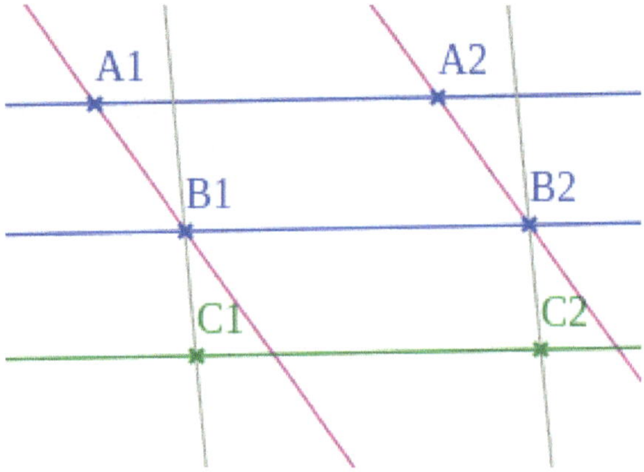

In affine geometry, one uses Playfair's axiom to find the line through C1 and parallel to B1B2, and to find the line through B2 and parallel to B1C1: their intersection C2 is the result of the indicated translation.

In 1748, Euler introduced the term *affine* (Latin *affinis*, "related") in his book *Introductio in analysin infinitorum* (volume 2, chapter XVIII). In 1827, August Möbius wrote on affine geometry in his *Der barycentrische Calcul* .

After Felix Klein›s Erlangen program, affine geometry was recognized as a generalization of Euclidean geometry.

In 1912, Edwin B. Wilson and Gilbert N. Lewis developed an affine geometry to express the special theory of relativity.

In 1918, Hermann Weyl referred to affine geometry for his text *Space, Time, Matter*. He used affine geometry to introduce vector addition and subtraction at the earliest stages of his development of mathematical physics. Later, E. T. Whittaker wrote:

Weyl's geometry is interesting historically as having been the first of the affine geometries to be worked out in detail: it is based on a special type of parallel transport ...using worldlines of light-signals in four-dimensional space-time. A short element of one of these world-lines may be called a *null-vector*; then the parallel transport in question is such that it carries any null-vector at one point into the position of a null-vector at a neighboring point.

In 1984, "the affine plane associated to the Lorentzian vector space L^2 " was described by Graciela Birman and Katsumi Nomizu in an article entitled "Trigonometry in Lorentzian geometry".

Systems of Axioms

Several axiomatic approaches to affine geometry have been put forward:

Pappus' Law

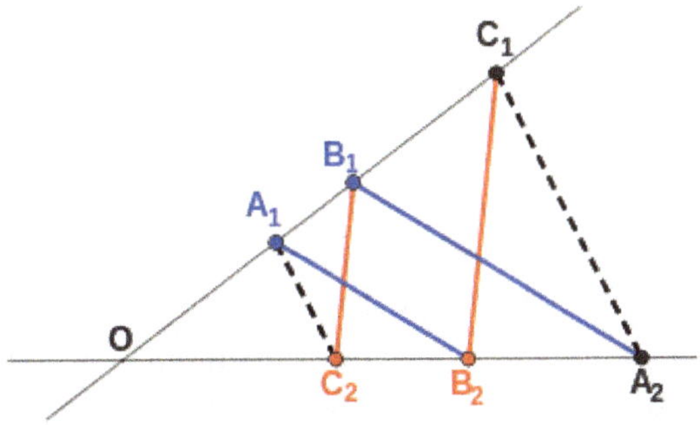

Pappus law: if the red lines are parallel and the blue lines are parallel, then the dotted black lines must be parallel.

As affine geometry deals with parallel lines, one of the properties of parallels noted by Pappus of Alexandria has been taken as a premise:

If A, B, C are on one line and A', B', C' on another, then

$(AB' \parallel A'B \wedge BC' \parallel B'C) \Rightarrow CA' \parallel C'A.$

The full axiom system proposed has *point*, *line*, and *line containing point* as primitive notions:

- Two points are contained in just one line.
- For any line *l* and any point *P*, not on *l*, there is just one line containing *P* and not containing any point of *l*. This line is said to be *parallel* to *l*.
- Every line contains at least two points.
- There are at least three points not belonging to one line.

According to H. S. M. Coxeter:

The interest of these five axioms is enhanced by the fact that they can be developed into a vast body of propositions, holding not only in Euclidean geometry but also in Minkowski's geometry of time and space (in the simple case of 1 + 1 dimensions, whereas the special theory of relativity needs 1 + 3). The extension to either Euclidean or Minkowskian geometry is achieved by adding various further axioms of orthogonality, etc

The various types of affine geometry correspond to what interpretation is taken for *rotation*. Euclidean geometry corresponds to the ordinary idea of rotation, while Minkowski's geometry corresponds to hyperbolic rotation. With respect to perpendicular lines, they remain perpendicular when the plane is subjected to ordinary rotation. In the Minkowski geometry, lines that are hyperbolic-orthogonal remain in that relation when the plane is subjected to hyperbolic rotation.

Ordered Structure

An axiomatic treatment of plane affine geometry can be built from the axioms of ordered geometry by the addition of two additional axioms:

- (Affine axiom of parallelism) Given a point A and a line r, not through A, there is at most one line through A which does not meet r.
- (Desargues) Given seven distinct points A, A', B, B', C, C', O, such that AA', BB', and CC' are distinct lines through O and AB is parallel to A'B' and BC is parallel to B'C', then AC is parallel to A'C'.

The affine concept of parallelism forms an equivalence relation on lines. Since the axioms of ordered geometry as presented here include properties that imply the structure of the real numbers, those properties carry over here so that this is an axiomatization of affine geometry over the field of real numbers.

Ternary Rings

The first non-Desarguesian plane was noted by David Hilbert in his *Foundations of Geometry*. The Moulton plane is a standard illustration. In order to provide a context for such geometry as well as those where Desargues theorem is valid, the concept of a ternary ring has been developed.

Rudimentary affine planes are constructed from ordered pairs taken from a ternary ring. A plane is said to have the "minor affine Desargues property" when two triangles in parallel perspective, having two parallel sides, must also have the third sides parallel. If this property holds in the rudimentary affine plane defined by a ternary ring, then there is an equivalence relation between "vectors" defined by pairs of points from the plane. Furthermore, the vectors form an abelian group under addition, the ternary ring is linear, and satisfies right distributivity:

$(a + b)\,c = ac + bc.$

Affine Transformations

Geometrically, affine transformations (affinities) preserve collinearity: so they transform parallel lines into parallel lines and preserve ratios of distances along parallel lines.

We identify as *affine theorems* any geometric result that is invariant under the affine group (in Felix Klein's Erlangen programme this is its underlying group of symmetry transformations for affine geometry). Consider in a vector space V, the general linear group GL(V). It is not the whole *affine group* because we must allow also translations by vectors v in V. (Such a translation maps any w in V to $w + v$.) The affine group is generated by the general linear group and the translations and is in fact their semidirect product $V \rtimes \mathrm{GL}(V)$. (Here we think of V as a group under its operation of addition, and use the defining representation of GL(V) on V to define the semidirect product.)

For example, the theorem from the plane geometry of triangles about the concurrence of the lines joining each vertex to the midpoint of the opposite side (at the *centroid* or *barycenter*) depends on the notions of *mid-point* and *centroid* as affine invariants. Other examples include the theorems of Ceva and Menelaus.

Affine invariants can also assist calculations. For example, the lines that divide the area of a triangle into two equal halves form an envelope inside the triangle. The ratio of the area of the envelope to the area of the triangle is affine invariant, and so only needs to be calculated from a simple case such as a unit isosceles right angled triangle to give $\frac{3}{4}\log_e(2) - \frac{1}{2}$, i.e. 0.019860... or less than 2%, for all triangles.

Familiar formulas such as half the base times the height for the area of a triangle, or a third the base times the height for the volume of a pyramid, are likewise affine invariants. While the latter is less obvious than the former for the general case, it is easily seen for the one-sixth of the unit cube formed by a face (area 1) and the midpoint of the cube (height 1/2). Hence it holds for all pyramids, even slanting ones whose apex is not directly above the center of the base, and those with base a parallelogram instead of a square. The formula further generalizes to pyramids whose base can be dissected into parallelograms, including cones by allowing infinitely many parallelograms (with due attention to convergence). The same approach shows that a four-dimensional pyramid has 4D volume one quarter the 3D volume of its parallelepiped base times the height, and so on for higher dimensions.

Affine Space

Affine geometry can be viewed as the geometry of an affine space of a given dimension n, coordinatized over a field K. There is also (in two dimensions) a combinatorial generalization of coordinatized affine space, as developed in synthetic finite geometry. In projective geometry, *affine space* means the complement of a hyperplane at infinity in a projective space. *Affine space* can also be viewed as a vector space whose operations are limited to those linear combinations whose coefficients sum to one, for example $2x - y$, $x - y + z$, $(x + y + z)/3$, $\mathbf{i}x + (1 - \mathbf{i})y$, etc.

Synthetically, affine planes are 2-dimensional affine geometries defined in terms of the relations between points and lines (or sometimes, in higher dimensions, hyperplanes). Defining affine (and projective) geometries as configurations of points and lines (or hyperplanes) instead of using coordinates, one gets examples with no coordinate fields. A major property is that all such examples have dimension 2. Finite examples in dimension 2 (finite affine planes) have been valuable in the study of configurations in infinite affine spaces, in group theory, and in combinatorics.

Despite being less general than the configurational approach, the other approaches discussed have been very successful in illuminating the parts of geometry that are related to symmetry.

Projective View

In traditional geometry, affine geometry is considered to be a study between Euclidean geometry and projective geometry. On the one hand, affine geometry is Euclidean geometry with congruence left out; on the other hand, affine geometry may be obtained from projective geometry by the designation of a particular line or plane to represent the points at infinity. In affine geometry, there is no metric structure but the parallel postulate does hold. Affine geometry provides the basis for Euclidean structure when perpendicular lines are defined, or the basis for Minkowski geometry through the notion of hyperbolic orthogonality. In this viewpoint, an affine transformation geometry is a group of projective transformations that do not permute finite points with points at infinity.

DISCRETE GEOMETRY

Discrete geometry and combinatorial geometry are branches of geometry that study combinatorial properties and constructive methods of discrete geometric objects. Most questions in discrete geometry involve finite or discrete sets of basic geometric objects, such as points, lines, planes, circles, spheres, polygons, and so forth. The subject focuses on the combinatorial properties of these objects, such as how they intersect one another, or how they may be arranged to cover a larger object.

Discrete geometry has large overlap with convex geometry and computational geometry, and is closely related to subjects such as finite geometry, combinatorial optimization, digital geometry, discrete differential geometry, geometric graph theory, toric geometry, and combinatorial topology.

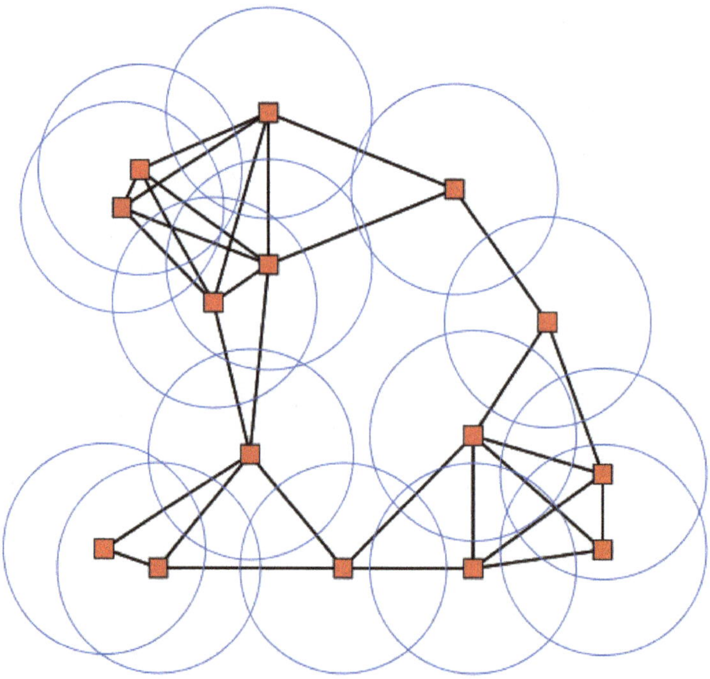

A collection of circles and the corresponding unit disk graph

Topics in Discrete Geometry

Polyhedra and Polytopes

A **polytope** is a geometric object with flat sides, which exists in any general number of dimensions. A polygon is a polytope in two dimensions, a polyhedron in three dimensions, and so on in higher dimensions (such as a 4-polytope in four dimensions). Some theories further generalize the idea to include such objects as unbounded polytopes (apeirotopes and tessellations), and abstract polytopes.

The following are some of the aspects of polytopes studied in discrete geometry:

- Polyhedral combinatorics
- Lattice polytopes
- Ehrhart polynomials
- Pick›s theorem
- Hirsch conjecture

Packings, Coverings and Tilings

Packings, coverings, and tilings are all ways of arranging uniform objects (typically circles, spheres, or tiles) in a regular way on a surface or manifold.

A **sphere packing** is an arrangement of non-overlapping spheres within a containing space. The spheres considered are usually all of identical size, and the space is usually three-dimensional Euclidean space. However, sphere packing problems can be generalised to consider unequal spheres, n-dimensional Euclidean space (where the problem becomes circle packing in two dimensions, or hypersphere packing in higher dimensions) or to non-Euclidean spaces such as hyperbolic space.

A **tessellation** of a flat surface is the tiling of a plane using one or more geometric shapes, called tiles, with no overlaps and no gaps. In mathematics, tessellations can be generalized to higher dimensions.

Specific topics in this area include:

- Circle packings
- Sphere packings
- Kepler conjecture
- Quasicrystals
- Aperiodic tilings
- Periodic Graphs (Geometry)
- Finite subdivision rules

Structural Rigidity and Flexibility

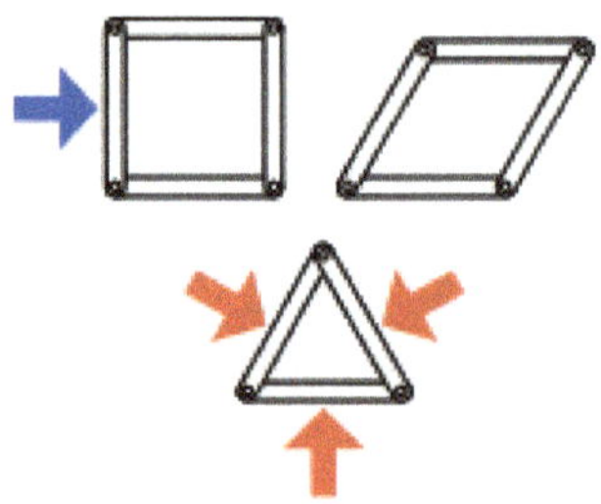

Graphs are drawn as rods connected by rotating hinges. The cycle graph C_4 drawn as a square can be tilted over by the blue force into a parallelogram, so it is a flexible graph. K_3, drawn as a triangle, cannot be altered by any force that is applied to it, so it is a rigid graph.

Structural rigidity is a combinatorial theory for predicting the flexibility of ensembles formed by rigid bodies connected by flexible linkages or hinges.

Topics in this area include:

- Cauchy›s theorem
- Flexible polyhedra

Incidence Structures

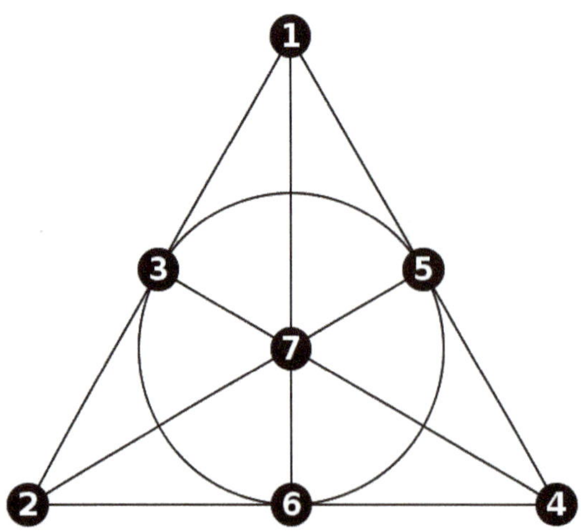

Seven points are elements of seven lines in the Fano plane, an example of an incidence structure.

Incidence structures generalize planes (such as affine, projective, and Möbius planes) as can be seen from their axiomatic definitions. Incidence structures also generalize the higher-dimensional analogs and the finite structures are sometimes called finite geometries.

Formally, an **incidence structure** is a triple

$$C = (P, L, I).$$

where P is a set of "points", L is a set of "lines" and $I \subseteq P \times L$ is the incidence relation. The elements of I are called **flags.** If

$$(p, l) \in I,$$

we say that point p "lies on" line I

Topics in this area include:

- Configurations
- Line arrangements
- Hyperplane arrangements
- Buildings

Oriented Matroids

An **oriented matroid** is a mathematical structure that abstracts the properties of directed graphs and of arrangements of vectors in a vector space over an ordered field (particularly for partially ordered vector spaces). In comparison, an ordinary (i.e., non-oriented) matroid abstracts the dependence properties that are common both to graphs, which are not necessarily *directed*, and to arrangements of vectors over fields, which are not necessarily *ordered*.

Geometric Graph Theory

A **geometric graph** is a graph in which the vertices or edges are associated with geometric objects. Examples include Euclidean graphs, the 1-skeleton of a polyhedron or polytope, intersection graphs, and visibility graphs.

Topics in this area include:

- Graph drawing
- Polyhedral graphs
- Voronoi diagrams and Delaunay triangulations

Simplicial Complexes

A **simplicial complex** is a topological space of a certain kind, constructed by "gluing together" points, line segments, triangles, and their n-dimensional counterparts (see illustration). Simplicial complexes should not be confused with the more abstract notion of a simplicial set appearing in modern simplicial homotopy theory. The purely combinatorial counterpart to a simplicial complex is an abstract simplicial complex.

Topological Combinatorics

The discipline of combinatorial topology used combinatorial concepts in topology and in the early 20th century this turned into the field of algebraic topology.

In 1978 the situation was reversed – methods from algebraic topology were used to solve a problem in combinatorics – when László Lovász proved the Kneser conjecture, thus beginning the new study of **topological combinatorics**. Lovász's proof used the Borsuk-Ulam theorem and this theorem retains a prominent role in this new field. This theorem has many equivalent versions and analogs and has been used in the study of fair division problems.

Topics in this are include:

- Sperner's lemma
- Regular maps

Lattices and Discrete Groups

A **discrete group** is a group *G* equipped with the discrete topology. With this topology, *G* becomes a topological group. A **discrete subgroup** of a topological group *G* is a subgroup *H* whose relative topology is the discrete one. For example, the integers, **Z**, form a discrete subgroup of the reals, **R** (with the standard metric topology), but the rational numbers, **Q**, do not.

A **lattice** in a locally compact topological group is a discrete subgroup with the property that the quotient space has finite invariant measure. In the special case of subgroups of $\mathbf{R}^n$, this amounts to the usual geometric notion of a lattice, and both the algebraic structure of lattices and the geometry of the totality of all lattices are relatively well understood. Deep results of Borel, Harish-Chandra, Mostow, Tamagawa, M. S. Raghunathan, Margulis, Zimmer obtained from the 1950s through the 1970s provided examples and generalized much of the theory to the setting of nilpotent Lie groups and semisimple algebraic groups over a local field. In the 1990s, Bass and Lubotzky initiated the study of *tree lattices*, which remains an active research area.

Topics in this area include:

- Reflection groups
- Triangle groups

Digital Geometry

Digital geometry deals with discrete sets (usually discrete point sets) considered to be digitized models or images of objects of the 2D or 3D Euclidean space.

Simply put, **digitizing** is replacing an object by a discrete set of its points. The images we see on the TV screen, the raster display of a computer, or in newspapers are in fact digital images.

Its main application areas are computer graphics and image analysis. See Li Chen, Digital and discrete geometry: Theory and Algorithms, Springer, 2014. (http://www.springer.com/us/book/9783319120980)

Discrete Differential Geometry

Discrete differential geometry is the study of discrete counterparts of notions in differential geometry. Instead of smooth curves and surfaces, there are polygons, meshes, and simplicial complexes. It is used in the study of computer graphics and topological combinatorics.

Topics in this area include:

- Discrete Laplace operator
- Discrete exterior calculus
- Discrete Morse theory
- Topological combinatorics
- Spectral shape analysis
- Abstract differential geometry
- Analysis on fractals

FINITE GEOMETRY

A finite geometry is any geometric system that has only a finite number of points. The familiar Euclidean geometry is not finite, because a Euclidean line contains infinitely many points. A geometry based on the graphics displayed on a computer screen, where the pixels are considered to be the points, would be a finite geometry. While there are many systems that could be called finite geometries, attention is mostly paid to the finite projective and affine spaces because of their regularity and simplicity. Other significant types of finite geometry are finite Möbius or inversive planes and Laguerre planes, which are examples of a general type called Benz planes, and their higher-dimensional analogs such as higher finite inversive geometries.

Finite geometries may be constructed via linear algebra, starting from vector spaces over a finite field; the affine and projective planes so constructed are called Galois geometries. Finite geometries can also be defined purely axiomatically. Most common finite geometries are Galois geometries, since any finite projective space of dimension three or greater is isomorphic to a projective space over a finite field (that is, the projectivization of a vector space over a finite field). However, dimension two has affine and projective planes that are not isomorphic to Galois geometries, namely the non-Desarguesian planes. Similar results hold for other kinds of finite geometries.

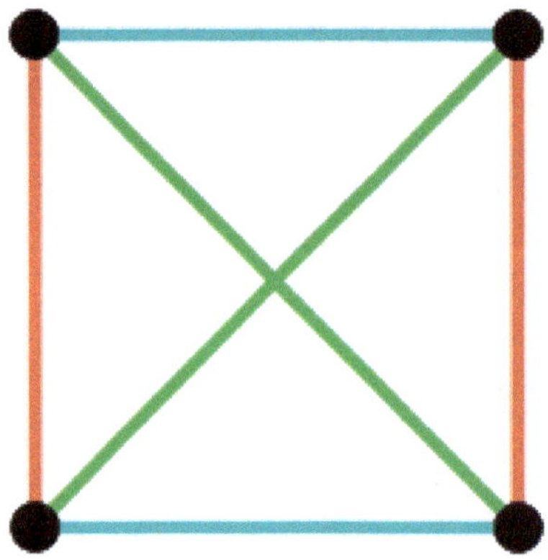

Finite affine plane of order 2, containing 4 points and 6 lines. Lines of the same color are "parallel".

Finite Affine and Projective Planes

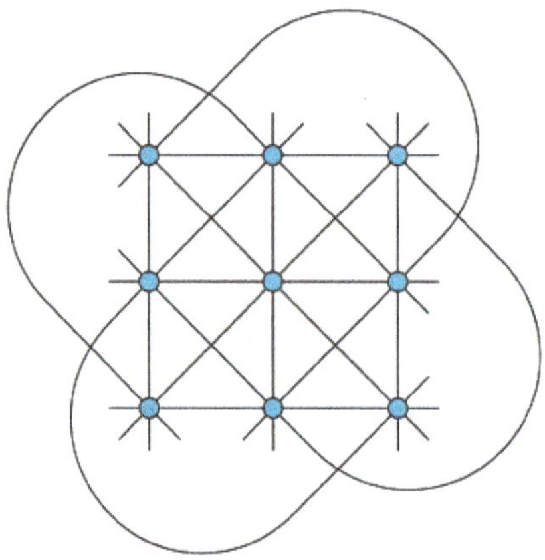

Finite affine plane of order 3, containing 9 points and 12 lines.

The following remarks apply only to finite *planes*. There are two main kinds of finite plane geometry: affine and projective. In an affine plane, the normal sense of parallel lines applies. In a projective plane, by contrast, any two lines intersect at a unique point, so parallel lines do not exist. Both finite affine plane geometry and finite projective plane geometry may be described by fairly simple axioms.

An affine plane geometry is a nonempty set X (whose elements are called "points"), along with a nonempty collection L of subsets of X (whose elements are called "lines"), such that:

- For every two distinct points, there is exactly one line that contains both points.
- Playfair›s axiom: Given a line ℓ; and a point p not on ℓ; there exists exactly one line ℓ; containing p such that $\ell \cap \ell' = \varnothing$.
- There exists a set of four points, no three of which belong to the same line.

The last axiom ensures that the geometry is not *trivial* (either empty or too simple to be of interest, such as a single line with an arbitrary number of points on it), while the first two specify the nature of the geometry.

The simplest affine plane contains only four points; it is called the *affine plane of order* 2. (The order of an affine plane is the number of points on any line, see below.) Since no three are collinear, any pair of points determines a unique line, and so this plane contains six lines. It corresponds to a tetrahedron where non-intersecting edges are considered "parallel", or a square where not only opposite sides, but also diagonals are considered "parallel". More generally, a finite affine plane of order n has n^2 points and $n^2 + n$ lines; each line contains n points, and each point is on $n + 1$ lines. The affine plane of order 3 is known as the Hesse configuration.

A projective plane geometry is a nonempty set X (whose elements are called "points"), along with a nonempty collection L of subsets of X (whose elements are called "lines"), such that:

- For every two distinct points, there is exactly one line that contains both points.
- The intersection of any two distinct lines contains exactly one point.
- There exists a set of four points, no three of which belong to the same line.

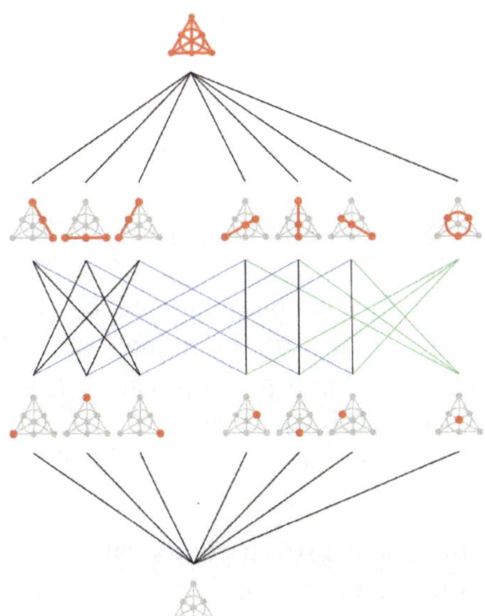

Duality in the Fano plane: Each point corresponds to a line and vice versa.

An examination of the first two axioms shows that they are nearly identical, except that the roles of points and lines have been interchanged. This suggests the principle of duality for projective plane geometries, meaning that any true statement valid in all these geometries remains true if we exchange points for lines and lines for points. The smallest geometry satisfying all three axioms contains seven points. In this simplest of the projective planes, there are also seven lines; each point is on three lines, and each line contains three points.

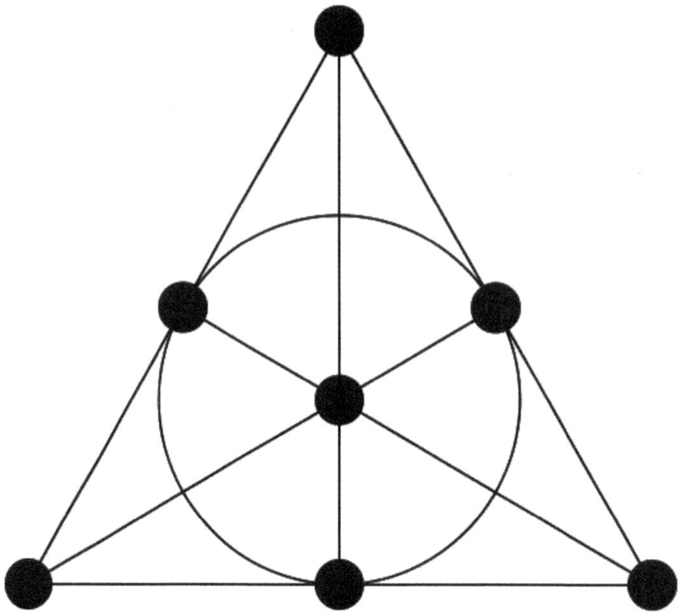

The Fano plane

This particular projective plane is sometimes called the **Fano plane**. If any of the lines is removed from the plane, along with the points on that line, the resulting geometry is the affine plane of order 2. The Fano plane is called **the** *projective plane of order* 2 because it is unique (up to isomorphism). In general, the projective plane of order n has $n^2 + n + 1$ points and the same number of lines; each line contains $n + 1$ points, and each point is on $n + 1$ lines.

A permutation of the Fano plane›s seven points that carries collinear points (points on the same line) to collinear points is called a collineation of the plane. The full collineation group is of order 168 and is isomorphic to the group PSL(2,7) ≈ PSL(3,2), which in this special case is also isomorphic to the general linear group GL(3,2) ≈ PGL(3,2).

Order of Planes

A finite plane of **order** n is one such that each line has n points (for an affine plane), or such that each line has $n + 1$ points (for a projective plane). One major open question in finite geometry is:

Is the order of a finite plane always a prime power?

This is conjectured to be true.

Affine and projective planes of order n exist whenever n is a prime power (a prime number raised to a positive integer exponent), by using affine and projective planes over the finite field with $n = p^k$ elements. Planes not derived from finite fields also exist, but all known examples have order a prime power.

The best general result to date is the Bruck–Ryser theorem of 1949, which states:

If n is a positive integer of the form $4k + 1$ or $4k + 2$ and n is not equal to the sum of two integer squares, then n does not occur as the order of a finite plane.

The smallest integer that is not a prime power and not covered by the Bruck–Ryser theorem is 10; 10 is of the form $4k + 2$, but it is equal to the sum of squares $1^2 + 3^2$. The non-existence of a finite plane of order 10 was proven in a computer-assisted proof that finished in 1989 – see (Lam 1991) for details.

The next smallest number to consider is 12, for which neither a positive nor a negative result has been proved.

Individual examples can be found in the work of Thomas Penyngton Kirkman (1847) and the systematic development of finite projective geometry given by von Staudt (1856).

The first axiomatic treatment of finite projective geometry was developed by the Italian mathematician Gino Fano. In his work on proving the independence of the set of axioms for projective n-space that he developed, he considered a finite three dimensional space with 15 points, 35 lines and 15 planes (see diagram), in which each line had only three points on it.

In 1906 Oswald Veblen and W. H. Bussey described projective geometry using homogeneous coordinates with entries from the Galois field GF(q). When $n + 1$ coordinates are used, the n-dimensional finite geometry is denoted PG(n, q). It arises in synthetic geometry and has an associated transformation group.

Finite Spaces of 3 or more Dimensions

For some important differences between finite *plane* geometry and the geometry of higher-dimensional finite spaces, see axiomatic projective space. For a discussion of higher-dimensional finite spaces in general, see, for instance, the works of J.W.P. Hirschfeld. The study of these higher-dimensional spaces ($n \geq 3$) has many important applications in advanced mathematical theories.

Axiomatic Definition

A **projective space** S can be defined axiomatically as a set P (the set of points), together with a set L of subsets of P (the set of lines), satisfying these axioms :

- Each two distinct points p and q are in exactly one line.
- Veblen›s axiom: If a, b, c, d are distinct points and the lines through ab and cd meet, then so do the lines through ac and bd.
- Any line has at least 3 points on it.

The last axiom eliminates reducible cases that can be written as a disjoint union of projective spaces together with 2-point lines joining any two points in distinct projective spaces. More abstractly, it can be defined as an incidence structure (P, L, I) consisting of a set P of points, a set L of lines, and an incidence relation I stating which points lie on which lines.

Obtaining a *finite* projective space requires one more axiom:

- The set of points P is a finite set.

In any finite projective space, each line contains the same number of points and the *order* of the space is defined as one less than this common number.

A subspace of the projective space is a subset X, such that any line containing two points of X is a subset of X (that is, completely contained in X). The full space and the empty space are always subspaces.

The *geometric dimension* of the space is said to be n if that is the largest number for which there is a strictly ascending chain of subspaces of this form:

$$\varnothing = X_{-1} \subset X_0 \subset \cdots \subset X_n = P.$$

Algebraic Construction

A standard algebraic construction of systems satisfies these axioms. For a division ring D construct an $(n + 1)$-dimensional vector space over D (vector space dimension is the number of elements in a basis). Let P be the 1-dimensional (single generator) subspaces and L the 2-dimensional (two independent generators) subspaces (closed under vector addition) of this vector space. Incidence is containment. If D is finite then it must be a finite field GF(q), since by Wedderburn's little theorem all finite division rings are fields. In this case, this construction produces a finite projective space. Furthermore, if the geometric dimension of a projective space is at least three then there is a division ring from which the space can be constructed in this manner. Consequently, all finite projective spaces of geometric dimension at least three are defined over finite fields. A finite projective space defined over such a finite field has $q + 1$ points on a line, so the two concepts of order coincide. Such a finite projective space is denoted by PG(n, q), where PG stands for projective geometry, n is the geometric dimension of the geometry and q is the size (order) of the finite field used to construct the geometry.

In general, the number k-dimensional subspaces of PG(n, q) is given by the product:

$$\binom{n+1}{k+1}_q = \prod_{i=0}^{k} \frac{q^{n+1-i}-1}{q^{i+1}-1},$$

which is a Gaussian binomial coefficient, a q analogue of a binomial coefficient.

Classification of finite projective spaces by geometric dimension

- Dimension 0 (no lines): The space is a single point and is so degenerate that it is usually ignored.
- Dimension 1 (exactly one line): All points lie on the unique line, called a *projective line*.
- Dimension 2: There are at least 2 lines, and any two lines meet. A projective space for $n = 2$ is a projective plane. These are much harder to classify, as not all of them are isomorphic with a PG(d, q). The Desarguesian planes (those that are isomorphic with a PG(2, q)) satisfy Desargues's theorem and are projective planes over finite fields, but there are many non-Desarguesian planes.
- Dimension at least 3: Two non-intersecting lines exist. The Veblen–Young theorem states in the finite case that every projective space of geometric dimension $n \geq 3$ is isomorphic with a PG(n, q), the n-dimensional projective space over some finite field GF(q).

The Smallest Projective Three-Space

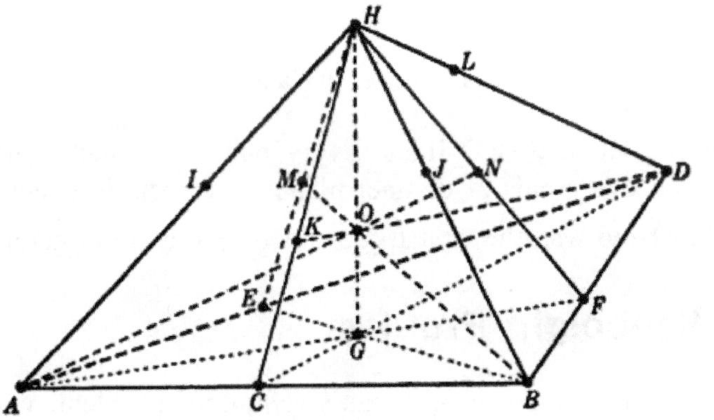

PG(3,2) but not all the lines are drawn

The smallest 3-dimensional projective space is over the field GF(2) and is denoted by PG(3,2). It has 15 points, 35 lines, and 15 planes. Each plane contains 7 points and 7 lines. Each line contains 3 points. As geometries, these planes are isomorphic to the Fano plane.

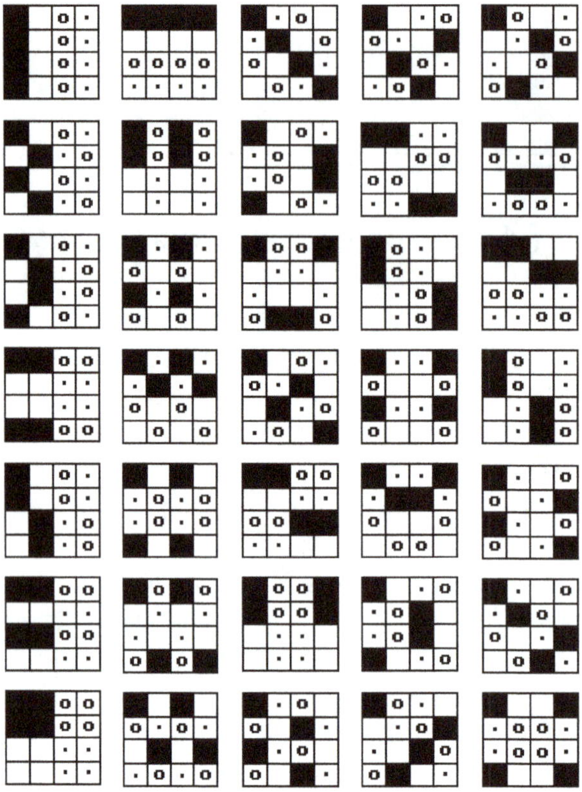

The square model of PG(3,2), the Fano 3-space.
Sets of black squares represent the 35 lines.
(This arrangement of the 35 lines illustrates
the Kirkman schoolgirls problem.)

Square model of Fano 3-space

Every point is contained in 7 lines. Every pair of distinct points are contained in exactly one line and every pair of distinct planes intersects in exactly one line.

In 1892, Gino Fano was the first to consider such a finite geometry.

Kirkman›s Schoolgirl Problem

PG(3,2) can arise as a solution of Kirkman's schoolgirl problem, which states: "Fifteen schoolgirls walk each day in five groups of three. Arrange the girls' walk for a week so that in that time, each pair of girls walks together in a group just once." (See answer in external link.) There are 35 different combinations for the girls to walk together. There are also 7 days of the week, and 3 girls in each group. Two of the seven non-isomorphic solutions to this problem can provide a visual representation of the Fano 3-space. Some diagrams for this problem can be found at :

Each color represents the day of the week (seven colors, blue, green, yellow, purple, red, black, and orange). The definition of a Fano space says that each line is on three points. The figure represents this showing that there are 3 points for every line. This is the basis for the answer to the schoolgirl problem. This figure is then rotated 7 times. There are 5 different lines for each day, multiplied by 7 (days) and the result is 35. Then, there are 15 points, and there are also 7 starting lines on each point. This then gives a representation of the Fano 3-space, PG(3,2).

REFERENCES

1. Artin, Emil (1988), Geometric algebra, Wiley Classics Library, New York: John Wiley & Sons Inc., pp. x+214, ISBN 0-471-60839-4, MR 1009557 (Reprint of the 1957 original; A Wiley-Interscience Publication)
2. Miller, Jeff. "Earliest Known Uses of Some of the Words of Mathematics (A)".
3. Blaschke, Wilhelm (1954). Analytische Geometrie. Basel: Birkhauser. p. 31.
4. Coxeter, H. S. M. (1969). Introduction to Geometry. New York: John Wiley & Sons. p. 191. ISBN 0-471-50458-0.
5. Edwin B. Wilson & Gilbert N. Lewis (1912). "The Space-time Manifold of Relativity. The Non-Euclidean Geometry of Mechanics and Electromagnetics", Proceedings of the American Academy of Arts and Sciences 48:387–507
6. Synthetic Spacetime, a digest of the axioms used, and theorems proved, by Wilson and Lewis. Archived by WebCite
7. Hermann Weyl (1918)Raum, Zeit, Materie. 5 edns. to 1922 ed. with notes by Jūrgen Ehlers, 1980. trans. 4th edn. Henry Brose, 1922 Space Time Matter, Methuen, rept. 1952 Dover. ISBN 0-486-60267-2 . See Chapter 1 §2 Foundations of Affine Geometry, pp 16–27
8. E. T. Whittaker (1958). From Euclid to Eddington: a study of conceptions of the external world, Dover Publications, p. 130.
9. Graciela S. Birman & Katsumi Nomizu (1984). "Trigonometry in Lorentzian geometry", American Mathematical Monthly 91(9):543–9, Lorentzian affine plane: p. 544
10. Veblen 1918: p. 103 (figure), and p. 118 (exercise 3).
11. Coxeter 1955, The Affine Plane, § 2: Affine geometry as an independent system
12. Coxeter 1955, Affine plane, p. 8
13. Coxeter, Introduction to Geometry, p. 192
14. David Hilbert, 1980 (1899). The Foundations of Geometry, 2nd ed., Chicago: Open Court, weblink from Project Gutenberg, p. 74.
15. Rafael Artzy (1965). Linear Geometry, Addison-Wesley, p. 213.
16. H. S. M. Coxeter (1942). Non-Euclidean Geometry, University of Toronto Press, pp. 18, 19.
17. Coxeter 1942, p. 178

CHAPTER 5

Zero / One-Dimensional Geometry

Chapter Contents

INTRODUCTION

In physics and mathematics, a sequence of n numbers can be understood as a location in n-dimensional space. When $n = 1$, the set of all such locations is called a **one-dimensional space**. An example of a one-dimensional space is the number line, where the position of each point on it can be described by a single number.

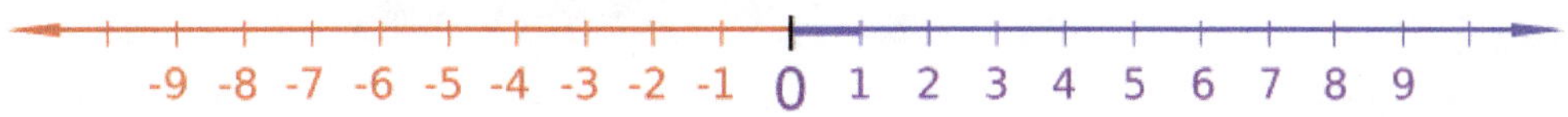

The number line

ONE-DIMENSIONAL GEOMETRY

Lines

The real number line and the real projective line are one-dimensional, though the latter is homeomorphic to a circle.

In algebraic geometry there are several structures which are technically one-dimensional spaces but referred to in other terms. For a field *k*, it is a one-dimensional vector space over itself. Similarly, the projective line over *k* is a one-dimensional space. In particular, if $k = \mathbb{C}$, the complex number plane, then the complex projective line $P^1(\mathbb{C})$ is one-dimensional with respect to $\mathbb{C}$, even if it is also known as the Riemann sphere.

More generally, a ring is a length-one module over itself. Similarly, the projective line over a ring is a one-dimensional space over the ring. In case the ring is an algebra over a field, these spaces are one-dimensional with respect to the algebra, even if the algebra is of higher dimensionality.

Polytopes

The only regular polytope in one dimension is the line segment, with the Schläfli symbol { }.

Hypersphere

The hypersphere in 1 dimension is a pair of points, sometimes called a **0-sphere** as its surface is zero-dimensional. Its length is

$$L = 2r$$

where r is the radius.

Coordinate Systems in One-Dimensional Space

The most popular coordinate systems are the number line and the angle.

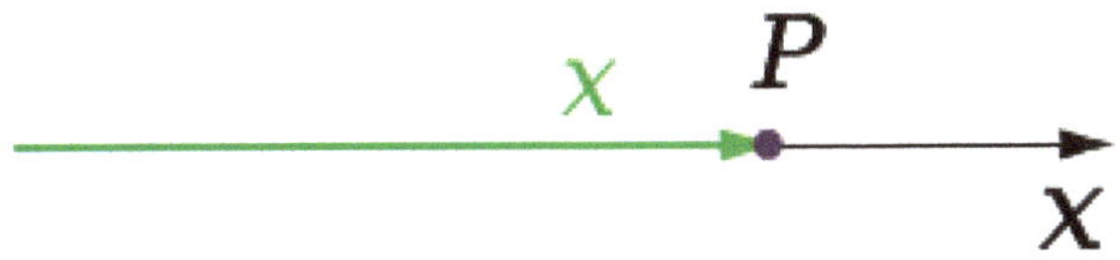

Number line

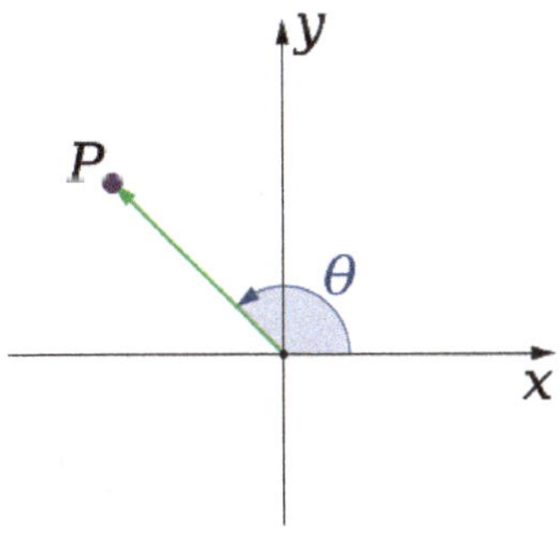

Angle

REFERENCES

1. "Пространство как математическое понятие» (in Russian). fmclass.ru. Retrieved 2015-06-06.
2. Gibilisco, Stan (1983). Understanding Einstein›s Theories of Relativity: Man›s New Perspective on the Cosmos. TAB Books. p. 89.

CHAPTER 6

Two-Dimensional Geometry

Chapter Contents

- **Introduction**

There are points and lines.

- 2 points determine a unique line, two lines are either parallel or determine a unique point which is in both lines. If the lines are parallel they have no point in common.
- Line segments, directed line segments, rays, angles triangles, are defined in the usual ways.
- There is a measure of length that:
    - Is additive on every line: the length of two disjoint or adjacent segments is the sum of their lengths.
    - Obeys the triangle inequality (any one side is shorter than the sum of the lengths of the other two).
- There is an additive measure of angle: so that the size of the sum of two angles is the sum of their sizes.
- Rays or line segments that make half of a flat (flat means a 180 degree angle) are normal, perpendicular and orthogonal to one another.

INTRODUCTION

In physics and mathematics, two-dimensional space or bi-dimensional space is a geometric model of the planar projection of the physical universe. The two dimensions are commonly called length and width. Both directions lie in the same plane.

A sequence of n real numbers can be understood as a location in n-dimensional space. When n = 2, the set of all such locations is called two-dimensional space or bi-dimensional space, and usually is thought of as a Euclidean space.

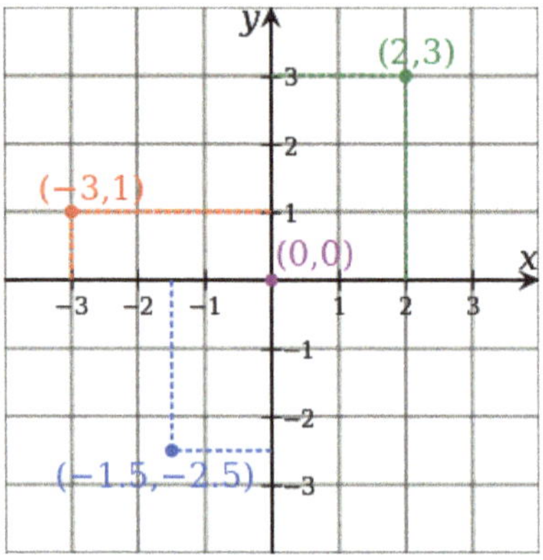

Bi-dimensional Cartesian coordinate system

Books I through IV and VI of Euclid's Elements dealt with two-dimensional geometry, developing such notions as similarity of shapes, the Pythagorean theorem (Proposition 47), equality of angles and areas, parallelism, the sum of the angles in a triangle, and the three cases in which triangles are "equal" (have the same area), among many other topics.

Later, the plane was described in a so-called *Cartesian coordinate system*, a coordinate system that specifies each point uniquely in a plane by a pair of numerical *coordinates*, which are the signed distances from the point to two fixed perpendicular directed lines, measured in the same unit of length. Each reference line is called a *coordinate axis* or just *axis* of the system, and the point where they meet is its *origin*, usually at ordered pair (0, 0). The coordinates can also be defined as the positions of the perpendicular projections of the point onto the two axes, expressed as signed distances from the origin.

The idea of this system was developed in 1637 in writings by Descartes and independently by Pierre de Fermat, although Fermat also worked in three dimensions, and did not publish the discovery. Both authors used a single axis in their treatments and have a variable length measured in reference to this axis. The concept of using a pair of axes was introduced later, after Descartes' *La Géométrie* was translated into Latin in 1649 by Frans van Schooten and his students. These commentators introduced several concepts while trying to clarify the ideas contained in Descartes' work.

Later, the plane was thought of as a field, where any two points could be multiplied and, except for 0, divided. This was known as the complex plane. The complex plane is sometimes called the Argand plane because it is used in Argand diagrams. These are named after Jean-Robert Argand (1768–1822), although they were first described by Norwegian-Danish land surveyor and mathematician Caspar Wessel (1745–1818).Argand diagrams are frequently used to plot the positions of the poles and zeroes of a function in the complex plane.

In Geometry

Coordinate Systems

"Plane coordinates" redirects here. It is not to be confused with Coordinate plane.

In mathematics, analytic geometry (also called Cartesian geometry) describes every point in two-dimensional space by means of two coordinates. Two perpendicular coordinate axes are given which cross each other at the origin. They are usually labeled x and y. Relative to these axes, the position of any point in two-dimensional space is given by an ordered pair of real numbers, each number giving the distance of that point from the origin measured along the given axis, which is equal to the distance of that point from the other axis.

Another widely used coordinate system is the polar coordinate system, which specifies a point in terms of its distance from the origin and its angle relative to a rightward reference ray.

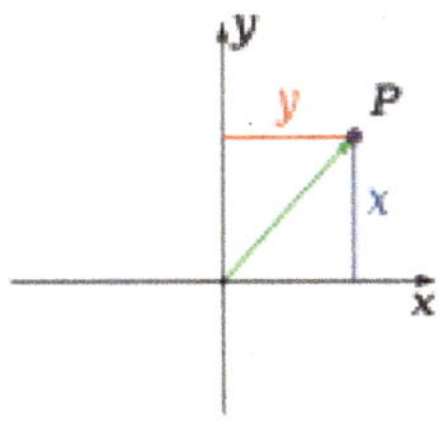

Cartesian coordinate system

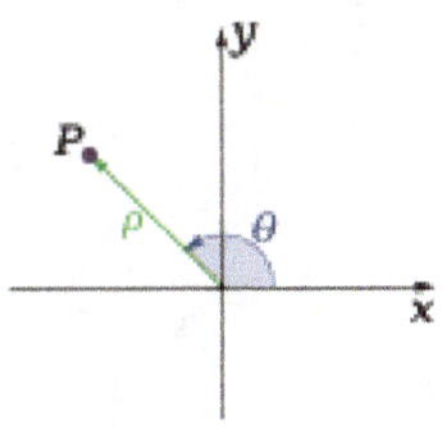

Polar coordinate system

Polytopes

In two dimensions, there are infinitely many polytopes: the polygons. The first few regular ones are shown below:

Convex

The Schläfli symbol {p} represents a regular *p*-gon.

Name	Triangle (2-simplex)	Square (2-orthoplex) (2-cube)	Pentagon	Hexagon	Heptagon	Octagon	
Schläfli	{3}	{4}	{5}	{6}	{7}	{8}	
Image	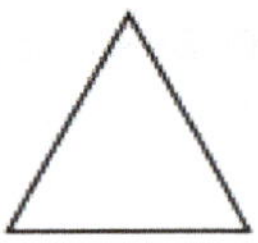		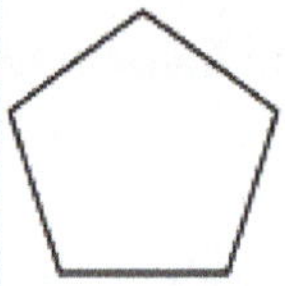	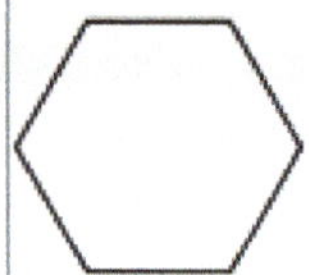	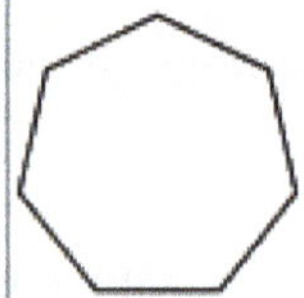	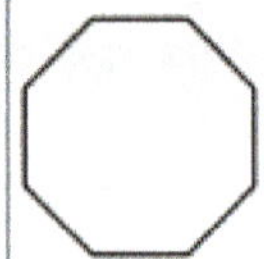	
Name	Nonagon	Decagon	Hendecagon	Dodecagon	Triskaidecagon	Tetradecagon	
Schläfli	{9}	{10}	{11}	{12}	{13}	{14}	
Image	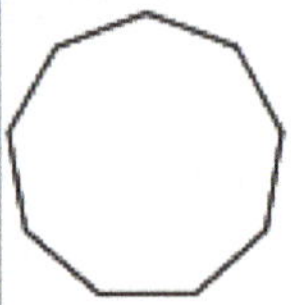	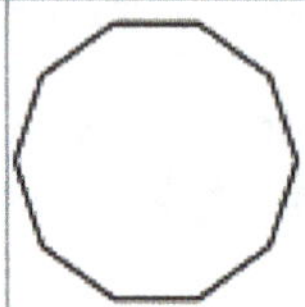	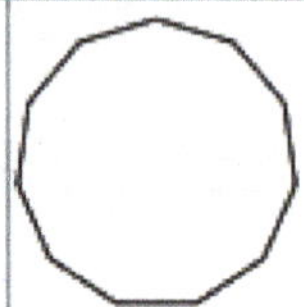	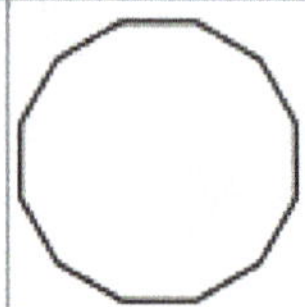	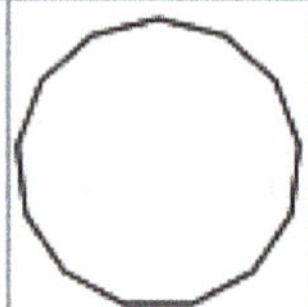	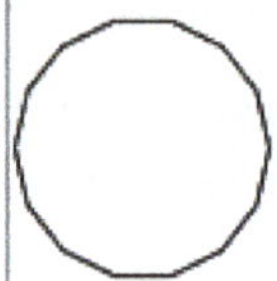	
Name	Pentadecagon	Hexadecagon	Heptadecagon	Octadecagon	Enneadecagon	Icosagon	...n-gon
Schläfli	{15}	{16}	{17}	{18}	{19}	{20}	{*n*}
Image	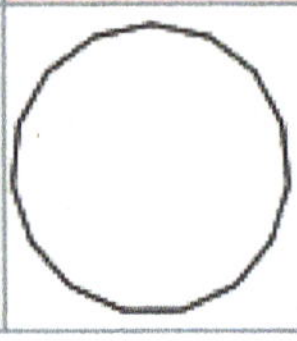	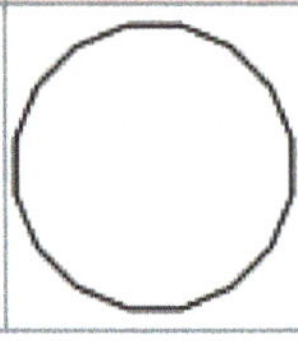	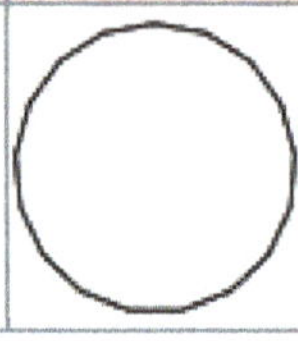	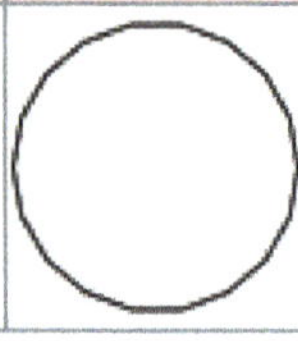	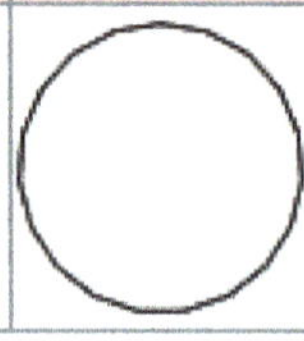	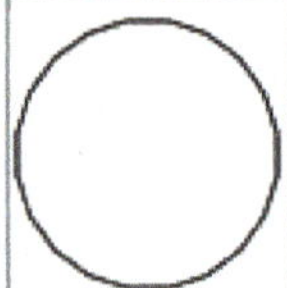	

Degenerate (Spherical)

The regular henagon {1} and regular digon {2} can be considered degenerate regular polygons. They can exist nondegenerately in non-Euclidean spaces like on a 2-sphere or a 2-torus.

Name	Henagon	Digon
Schläfli	{1}	{2}
Image	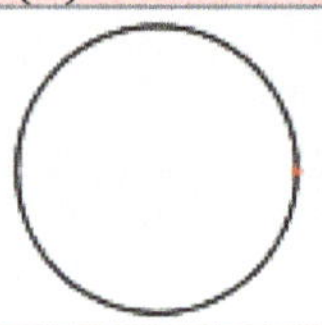	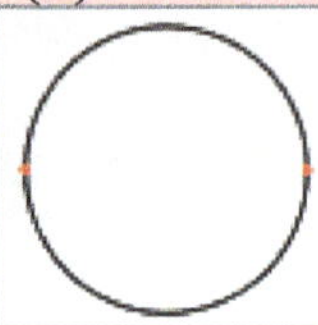

Non-Convex

There exist infinitely many non-convex regular polytopes in two dimensions, whose Schläfli symbols consist of rational numbers {n/m}. They are called star polygons and share the same vertex arrangements of the convex regular polygons.

In general, for any natural number n, there are n-pointed non-convex regular polygonal stars with Schläfli symbols $\{n/m\}$ for all m such that $m < n/2$ (strictly speaking $\{n/m\} = \{n/(n-m)\}$) and m and n are coprime.

Name	Pentagram	Heptagrams		Octagram	Enneagrams		Decagram	...n-agrams
Schläfli	{5/2}	{7/2}	{7/3}	{8/3}	{9/2}	{9/4}	{10/3}	$\{n/m\}$
Image		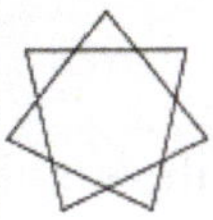		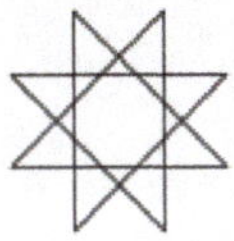	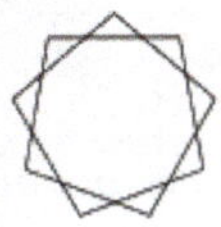		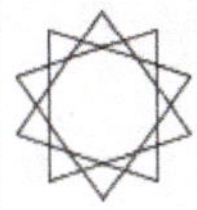	

Circle

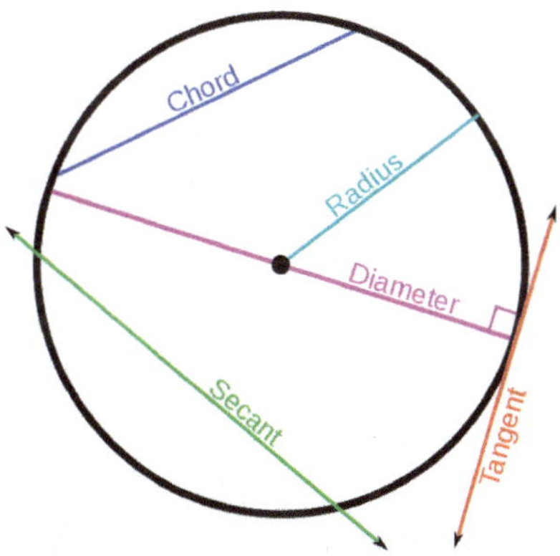

The hypersphere in 2 dimensions is a circle, sometimes called a 1-sphere (S^1) because it is a one-dimensional manifold. In a Euclidean plane, it has the length $2\pi r$ and the area of its interior is

$$A = \pi r^2$$

where r is the radius.

Other Shapes

There are an infinitude of other curved shapes in two dimensions, notably including the conic sections: the ellipse, the parabola, and the hyperbola.

In Linear Algebra

Another mathematical way of viewing two-dimensional space is found in linear algebra, where the idea of independence is crucial. The plane has two dimensions because the length of a rectangle is independent of its width. In the technical language of linear algebra, the plane is two-dimensional because every point in the plane can be described by a linear combination of two independent vectors.

Dot Product, Angle, and Length

The dot product of two vectors $\mathbf{A} = [A_1, A_2]$ and $\mathbf{B} = [B_1, B_2]$ is defined as:

$$\mathbf{A} \cdot \mathbf{B} = A_1 B_1 + A_2 B_2$$

A vector can be pictured as an arrow. Its magnitude is its length, and its direction is the direction the arrow points. The magnitude of a vector **A** is denoted by $\|\mathbf{A}\|$. In this viewpoint, the dot product of two Euclidean vectors **A** and **B** is defined by

$$\mathbf{A} \cdot \mathbf{B} = \|\mathbf{A}\| \, \|\mathbf{B}\| \cos \theta,$$

where θ is the angle between **A** and **B**.

The dot product of a vector **A** by itself is

$$\mathbf{A} \cdot \mathbf{A} = \|\mathbf{A}\|^2,$$

which gives

$$\|\mathbf{A}\| = \sqrt{\mathbf{A} \cdot \mathbf{A}},$$

the formula for the Euclidean length of the vector.

In Calculus

Gradient

In a rectangular coordinate system, the gradient is given by

$$\nabla f = \frac{\partial f}{\partial x}\mathbf{i} + \frac{\partial f}{\partial y}\mathbf{j}$$

Line Integrals and Double Integrals

For some scalar field $f : U \subseteq \mathbf{R}^2 \to \mathbf{R}$, the line integral along a piecewise smooth curve $C \subset U$ is defined as

$$\int\limits_C f\, ds = \int_a^b f(\mathbf{r}(t))|\mathbf{r}'(t)|\, dt.$$

where **r**: [a, b] → C is an arbitrary bijective parametrization of the curve C such that **r**(a) and **r**(b) give the endpoints of C and $a < b$. For a vector field $\mathbf{F} : U \subseteq \mathbf{R}^2 \to \mathbf{R}^2$, the line integral along a piecewise smooth curve $C \subset U$, in the direction of **r**, is defined as

$$\int\limits_C \mathbf{F}(\mathbf{r}) \cdot d\mathbf{r} = \int_a^b \mathbf{F}(\mathbf{r}(t)) \cdot \mathbf{r}'(t)\, dt.$$

where · is the dot product and **r**: [a, b] → C is a bijective parametrization of the curve C such that **r**(a) and **r**(b) give the endpoints of C.

A double integral refers to an integral within a region D in $\mathbf{R}^2$ of a function $f(x, y)$ and is usually written as:

$$\iint\limits_D f(x, y)\, dx\, dy.$$

Fundamental Theorem of Line Integrals

The fundamental theorem of line integrals, says that a line integral through a gradient field can be evaluated by evaluating the original scalar field at the endpoints of the curve. Let $\varphi : U \subseteq \mathbb{R}^2 \to \mathbb{R}$. Then

$$\varphi(\mathbf{q}) - \varphi(\mathbf{p}) = \int_{\gamma[\mathbf{p},\, \mathbf{q}]} \nabla\varphi(\mathbf{r}) \cdot d\mathbf{r}.$$

Green's Theorem

Let C be a positively oriented, piecewise smooth, simple closed curve in a plane, and let D be the region bounded by C. If L and M are functions of (x, y) defined on an open region containing D and have continuous partial derivatives there, then

$$\oint_C (L\, dx + M\, dy) = \iint_D \left(\frac{\partial M}{\partial x} - \frac{\partial L}{\partial y}\right) dx\, dy$$

where the path of integration along C is counterclockwise.

In Topology

In topology, the plane is characterized as being the unique contractible 2-manifold.

Its dimension is characterized by the fact that removing a point from the plane leaves a space that is connected, but not simply connected.

In Graph Theory

In graph theory, a planar graph is a graph that can be embedded in the plane, i.e., it can be drawn on the plane in such a way that its edges intersect only at their endpoints. In other words, it can be drawn in such a way that no edges cross each other. Such a drawing is called a *plane graph* or *planar embedding of the graph*. A plane graph can be defined as a planar graph with a mapping from every node to a point on a plane, and from every edge to a plane curve on that plane, such that the extreme points of each curve are the points mapped from its end nodes, and all curves are disjoint except on their extreme points.

REFERENCES

1. "Analytic geometry". Encyclopædia Britannica (Encyclopædia Britannica Online ed.). 2008.
2. Burton 2011, p. 374
3. Wessel's memoir was presented to the Danish Academy in 1797; Argand's paper was published in 1806. (Whittaker & Watson, 1927, p. 9)
4. S. Lipschutz; M. Lipson (2009). Linear Algebra (Schaum's Outlines) (4th ed.). McGraw Hill. ISBN 978-0-07-154352-1.
5. M.R. Spiegel; S. Lipschutz; D. Spellman (2009). Vector Analysis (Schaum's Outlines) (2nd ed.). McGraw Hill. ISBN 978-0-07-161545-7.
6. Mathematical methods for physics and engineering, K.F. Riley, M.P. Hobson, S.J. Bence, Cambridge University Press, 2010, ISBN 978-0-521-86153-3
7. Vector Analysis (2nd Edition), M.R. Spiegel, S. Lipschutz, D. Spellman, Schaum's Outlines, McGraw Hill (USA), 2009, ISBN 978-0-07-161545-7
8. Trudeau, Richard J. (1993). Introduction to Graph Theory (Corrected, enlarged republication. ed.). New York: Dover Pub. p. 64. ISBN 978-0-486-67870-2. Retrieved 8 August 2012. Thus a planar graph, when drawn on a flat surface, either has no edge-crossings or can be redrawn without them.

CHAPTER 7

Three-Dimensional Geometry

Chapter Contents

The question of why space is three-dimensional (3D) and not some other number of dimensions has puzzled philosophers and scientists since ancient Greece. Space-time overall is four-dimensional, or (3 + 1)-dimensional, where time is the fourth dimension. It's well-known that the time dimension is related to the second law of thermodynamics: time has one direction (forward) because entropy (a measure of disorder) never decreases in a closed system such as the universe.

INTRODUCTION

Three-dimensional space (also: 3-space or, rarely, tri-dimensional space) is a geometric setting in which three values (called parameters) are required to determine the position of an element (i.e., point). This is the informal meaning of the term dimension.

In physics and mathematics, a sequence of n numbers can be understood as a location in n-dimensional space. When n = 3, the set of all such locations is called three-dimensional Euclidean space. It is commonly represented by the symbol ℝ3. This serves as a three-parameter model of the physical universe (that is, the spatial part, without considering time) in which all known matter exists. However, this space is only one example of a large variety of spaces in three dimensions called 3-manifolds. In this classical example, when the three values refer to measurements in different directions (coordinates), any three directions can be chosen, provided that vectors in these directions do not all lie in the same 2-space (plane). Furthermore, in this case, these three values can be labeled by any combination of three chosen from the terms width, height, depth, and breadth.

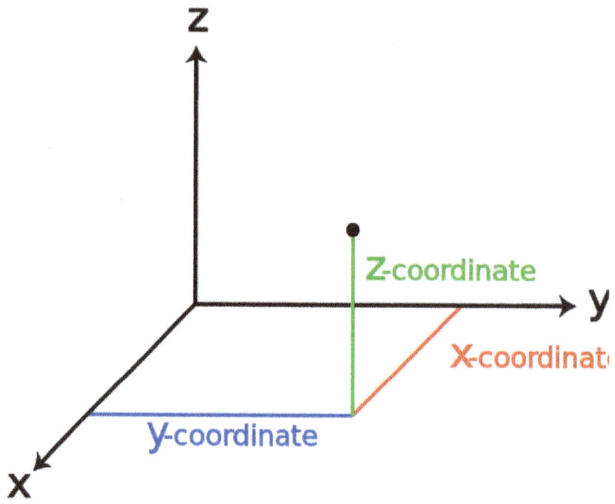

A representation of a three-dimensional Cartesian coordinate system with the x-axis pointing towards the observer.

In Euclidean Geometry

Coordinate Systems

In mathematics, analytic geometry (also called Cartesian geometry) describes every point in three-dimensional space by means of three coordinates. Three coordinate axes are given, each perpendicular to the other two at the origin, the point at which they cross. They are usually labeled x, y, and z. Relative to these axes, the position of any point in three-dimensional space is given by an ordered triple of real numbers, each number giving the distance of that point from the origin measured along the given axis, which is equal to the distance of that point from the plane determined by the other two axes.

Other popular methods of describing the location of a point in three-dimensional space include cylindrical coordinates and spherical coordinates, though there are an infinite number of possible methods. See Euclidean space.

Below are images of the above-mentioned systems.

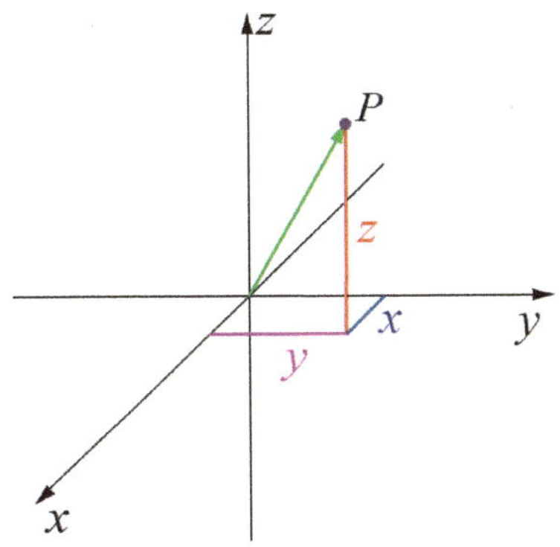

Cartesian coordinate system

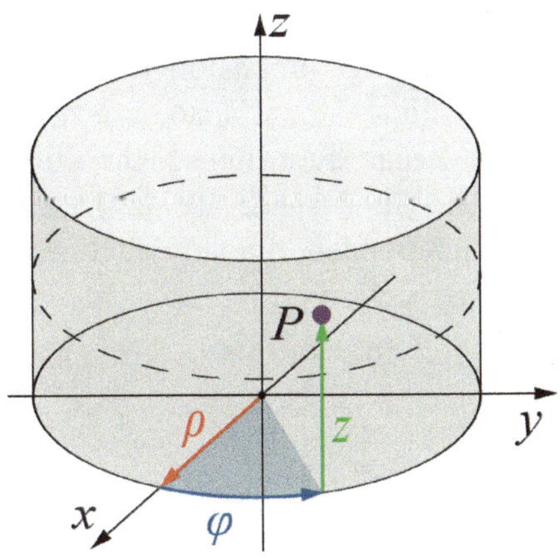

Cylindrical coordinate system

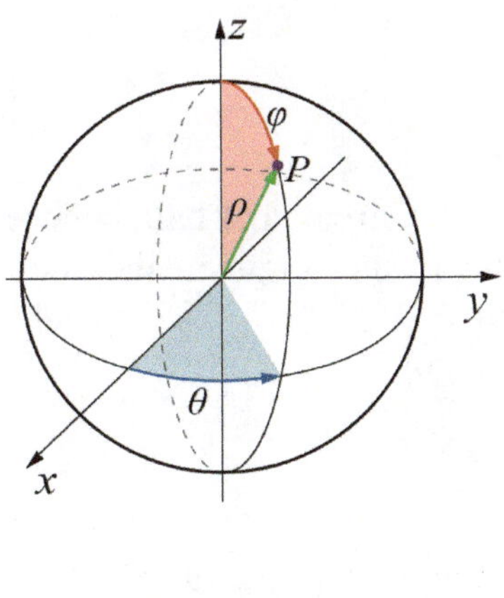

Spherical coordinate system

Lines and Planes

Two distinct points always determine a (straight) line. Three distinct points are either collinear or determine a unique plane. Four distinct points can either be collinear, coplanar or determine the entire space.

Two distinct lines can either intersect, be parallel or be skew. Two parallel lines, or two intersecting lines, lie in a unique plane, so skew lines are lines that do not meet and do not lie in a common plane.

Two distinct planes can either meet in a common line or are parallel (do not meet). Three distinct planes, no pair of which are parallel, can either meet in a common line, meet in a unique common point or have no point in common. In the last case, the three lines of intersection of each pair of planes are mutually parallel.

A line can lie in a given plane, intersect that plane in a unique point or be parallel to the plane. In the last case, there will be lines in the plane that are parallel to the given line.

A hyperplane is a subspace of one dimension less than the dimension of the full space. The hyperplanes of a three-dimensional space are the two-dimensional subspaces, that is, the planes. In terms of cartesian coordinates, the points of a hyperplane satisfy a single linear equation, so planes in this 3-space are described by linear equations. A line can be described by a pair of independent linear equations, each representing a plane having this line as a common intersection.

Varignon's theorem states that the midpoints of any quadrilateral in $\mathbb{R}^3$ form a parallelogram, and so, are coplanar.

Spheres and Balls

A sphere in 3-space (also called a **2-sphere** because it is a 2-dimensional object) consists of the set of all points in 3-space at a fixed distance r from a central point P. The solid enclosed by the sphere is called a **ball** (or, more precisely a **3-ball**). The volume of the ball is given by

$$V = \frac{4}{3}\pi r^3.$$

Another type of sphere arises from a 4-ball, whose three-dimensional surface is the **3-sphere**: points equidistant to the origin of the euclidean space $\mathbb{R}^4$. If a point has coordinates, $P(x, y, z, w)$, then $x^2 + y^2 + z^2 + w^2 = 1$ characterizes those points on the unit 3-sphere centered at the origin.

Polytopes

In three dimensions, there are nine regular polytopes: the five convex Platonic solids and the four nonconvex Kepler-Poinsot polyhedra.

Regular polytopes in three dimensions									
Class	Platonic solids					Kepler-Poinsot polyhedra			
Symmetry	T_d	O_h		I_h					
Coxeter group	A_3, [3,3]	B_3, [4,3]		H_3, [5,3]					
Order	24	48		120					
Regular polyhedron	 {3,3}	 {4,3}	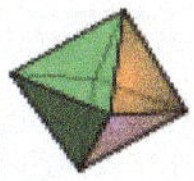 {3,4}	 {5,3}	 {3,5}	 {5/2,5}	 {5,5/2}	 {5/2,3}	 {3,5/2}

Surfaces of Revolution

A surface generated by revolving a plane curve about a fixed line in its plane as an axis is called a surface of revolution. The plane curve is called the *generatrix* of the surface. A section of the surface, made by intersecting the surface with a plane that is perpendicular (orthogonal) to the axis, is a circle.

Simple examples occur when the generatrix is a line. If the generatrix line intersects the axis line, the surface of revolution is a right circular cone with vertex (apex) the point of intersection. However, if the generatrix and axis are parallel, the surface of revolution is a circular cylinder.

Quadric Surfaces

In analogy with the conic sections, the set of points whose cartesian coordinates satisfy the general equation of the second degree, namely,

$$Ax^2 + By^2 + Cz^2 + Fxy + Gyz + Hxz + Jx + Ky + Lz + M = 0,$$

where A, B, C, F, G, H, J, K, L and M are real numbers and not all of A, B, C, F, G and H are zero is called a **quadric surface**.

There are six types of non-degenerate quadric surfaces:

- Ellipsoid
- Hyperboloid of one sheet
- Hyperboloid of two sheets
- Elliptic cone
- Elliptic paraboloid
- Hyperbolic paraboloid

The degenerate quadric surfaces are the empty set, a single point, a single line, a single plane, a pair of planes or a quadratic cylinder (a surface consisting of a non-degenerate conic section in a plane π and all the lines of $\mathbb{R}^3$ through that conic that are normal to π). Elliptic cones are sometimes considered to be degenerate quadric surfaces as well.

Both the hyperboloid of one sheet and the hyperbolic paraboloid are ruled surfaces, meaning that they can be made up from a family of straight lines. In fact, each has two families of generating lines, the members of each family are disjoint and each member one family intersects, with just one exception, every member of the other family.

In Linear Algebra

Another way of viewing three-dimensional space is found in linear algebra, where the idea of independence is crucial. Space has three dimensions because the length of a box is independent of its width or breadth. In the technical language of linear algebra, space is three-dimensional because every point in space can be described by a linear combination of three independent vectors.

Dot product, Angle, and Length

A vector can be pictured as an arrow. Its magnitude is its length, and its direction is the direction the arrow points. A vector in $\mathbb{R}^3$ can be represented by an ordered triple of real numbers. These numbers are called the **components** of the vector.

The dot product of two vectors $\mathbf{A} = [A_1, A_2, A_3]$ and $\mathbf{B} = [B_1, B_2, B_3]$ is defined as:

$$\mathbf{A} \cdot \mathbf{B} = A_1B_1 + A_2B_2 + A_3B_3.$$

The magnitude of a vector $\mathbf{A}$ is denoted by $\|\mathbf{A}\|$. The dot product of a vector $\mathbf{A} = [A_1, A_2, A_3]$ with itself is

$$\mathbf{A} \cdot \mathbf{A} = \|\mathbf{A}\|^2 = A_1^2 + A_2^2 + A_3^2,$$

which gives

$$\|\mathbf{A}\| = \sqrt{\mathbf{A} \cdot \mathbf{A}} = \sqrt{A_1^2 + A_2^2 + A_3^2},$$

the formula for the Euclidean length of the vector.

Without reference to the components of the vectors, the dot product of two non-zero Euclidean vectors $\mathbf{A}$ and $\mathbf{B}$ is given by

$$\mathbf{A} \cdot \mathbf{B} = \|\mathbf{A}\| \, \|\mathbf{B}\| \cos \theta,$$

where θ is the angle between $\mathbf{A}$ and $\mathbf{B}$.

Cross Product

The cross product or **vector product** is a binary operation on two vectors in three-dimensional space and is denoted by the symbol $\times$. The cross product $\mathbf{a} \times \mathbf{b}$ of the vectors $\mathbf{a}$ and $\mathbf{b}$ is a vector that is perpendicular to both and therefore normal to the plane containing them. It has many applications in mathematics, physics, and engineering.

The space and product form an algebra over a field, which is neither commutative nor associative, but is a Lie algebra with the cross product being the Lie bracket.

One can in n dimensions take the product of $n - 1$ vectors to produce a vector perpendicular to all of them. But if the product is limited to non-trivial binary products with vector results, it exists only in three and seven dimensions.

In Calculus

Gradient, Divergence and Curl

In a rectangular coordinate system, the gradient is given by

$$\nabla f = \frac{\partial f}{\partial x}\mathbf{i} + \frac{\partial f}{\partial y}\mathbf{j} + \frac{\partial f}{\partial z}\mathbf{k}$$

The divergence of a continuously differentiable vector field $\mathbf{F} = U\,\mathbf{i} + V\,\mathbf{j} + W\,\mathbf{k}$ is equal to the scalar-valued function:

$$\operatorname{div}\mathbf{F} = \nabla\cdot\mathbf{F} = \frac{\partial U}{\partial x} + \frac{\partial V}{\partial y} + \frac{\partial W}{\partial z}.$$

Expanded in Cartesian coordinates (see Del in cylindrical and spherical coordinates for spherical and cylindrical coordinate representations), the curl $\nabla \times \mathbf{F}$ is, for $\mathbf{F}$ composed of $[F_x, F_y, F_z]$:

$$\begin{vmatrix} \mathbf{i} & \mathbf{j} & \mathbf{k} \\ \frac{\partial}{\partial x} & \frac{\partial}{\partial y} & \frac{\partial}{\partial z} \\ F_x & F_y & F_z \end{vmatrix}$$

where **i**, **j**, and **k** are the unit vectors for the *x*-, *y*-, and *z*-axes, respectively. This expands as follows:

$$\left(\frac{\partial F_z}{\partial y} - \frac{\partial F_y}{\partial z}\right)\mathbf{i} + \left(\frac{\partial F_x}{\partial z} - \frac{\partial F_z}{\partial x}\right)\mathbf{j} + \left(\frac{\partial F_y}{\partial x} - \frac{\partial F_x}{\partial y}\right)\mathbf{k}$$

Line Integrals, Surface Integrals, and Volume Integrals

For some scalar field $f : U \subseteq \mathbf{R}^n \to \mathbf{R}$, the line integral along a piecewise smooth curve $C \subset U$ is defined as

$$\int_C f\,ds = \int_a^b f(\mathbf{r}(t))|\mathbf{r}'(t)|\,dt.$$

where $\mathbf{r}$: [a, b] $\to C$ is an arbitrary bijective parametrization of the curve C such that $\mathbf{r}(a)$ and $\mathbf{r}(b)$ give the endpoints of C and $a < b$ For a vector field $\mathbf{F} : U \subseteq \mathbf{R}^n \to \mathbf{R}^n$, the line integral along a piecewise smooth curve $C \subset U$, in the direction of $\mathbf{r}$, is defined as

$$\int_C \mathbf{F}(\mathbf{r}) \cdot d\mathbf{r} = \int_a^b \mathbf{F}(\mathbf{r}(t)) \cdot \mathbf{r}'(t)\, dt.$$

where $\cdot$ is the dot product and $\mathbf{r}$: [a, b] $\to C$ is a bijective parametrization of the curve C such that $\mathbf{r}(a)$ and $\mathbf{r}(b)$ give the endpoints of C.

A surface integral is a generalization of multiple integrals to integration over surfaces. It can be thought of as the double integral analog of the line integral. To find an explicit formula for the surface integral, we need to parameterize the surface of interest, S, by considering a system of curvilinear coordinates on S, like the latitude and longitude on a sphere. Let such a parameterization be $\mathbf{x}(s, t)$, where (s, t) varies in some region T in the plane. Then, the surface integral is given by

$$\iint_S f\, dS = \iint_T f(\mathbf{x}(s,t)) \left\| \frac{\partial \mathbf{x}}{\partial s} \times \frac{\partial \mathbf{x}}{\partial t} \right\| ds\, dt$$

where the expression between bars on the right-hand side is the magnitude of the cross product of the partial derivatives of $\mathbf{x}(s, t)$, and is known as the surface element. Given a vector field $\mathbf{v}$ on S, that is a function that assigns to each $\mathbf{x}$ in S a vector $\mathbf{v}(\mathbf{x})$, the surface integral can be defined component-wise according to the definition of the surface integral of a scalar field; the result is a vector.

A volume integral refers to an integral over a 3-dimensional domain.

It can also mean a triple integral within a region D in $\mathbf{R}^3$ of a function $f(x, y, z)$, and is usually written as:

$$\iiint_D f(x, y, z)\, dx\, dy\, dz.$$

Fundamental Theorem of Line Integrals

The fundamental theorem of line integrals, says that a line integral through a gradient field can be evaluated by evaluating the original scalar field at the endpoints of the curve.

Let $\varphi : U \subseteq \mathbb{R}^n \to \mathbb{R}$. Then

$$\varphi(\mathbf{q}) - \varphi(\mathbf{p}) = \int_{\gamma[\mathbf{p},\, \mathbf{q}]} \nabla\varphi(\mathbf{r}) \cdot d\mathbf{r}.$$

Stokes' Theorem

Stokes' theorem relates the surface integral of the curl of a vector field F over a surface Σ in Euclidean three-space to the line integral of the vector field over its boundary $\partial\Sigma$:

$$\iint_{\Sigma} \nabla \times \mathbf{F} \cdot d\mathbf{\Sigma} = \oint_{\partial\Sigma} \mathbf{F} \cdot d\mathbf{r}.$$

Divergence Theorem

Suppose V is a subset of $\mathbb{R}^n$ (in the case of $n = 3$, V represents a volume in 3D space) which is compact and has a piecewise smooth boundary S(also indicated with $\partial V = S$). If $\mathbf{F}$ is a continuously differentiable vector field defined on a neighborhood of V, then the divergence theorem says:

$$\iiint_V (\nabla \cdot \mathbf{F})\, dV = \oiint_S (\mathbf{F} \cdot \mathbf{n})\, dS.$$

The left side is a volume integral over the volume V, the right side is the surface integral over the boundary of the volume V. The closed manifold ∂V is quite generally the boundary of V oriented by outward-pointing normals, and $\mathbf{n}$ is the outward pointing unit normal field of the boundary ∂V. ($d\mathbf{S}$ may be used as a shorthand for $\mathbf{n}dS$.)

In topology

Three-dimensional space has a number of topological properties that distinguish it from spaces of other dimension numbers. For example, at least three dimensions are required to tie a knot in a piece of string.

With the space R^3, the topologists locally model all other 3-manifolds.

REFERENCES

1. Anton, Howard (1994), Elementary Linear Algebra (7th ed.), John Wiley & Sons, ISBN 978-0-471-58742-2
2. Arfken, George B. and Hans J. Weber. Mathematical Methods For Physicists, Academic Press; 6 edition (June 21, 2005). ISBN 978-0-12-059876-2.
3. Brannan, David A.; Esplen, Matthew F.; Gray, Jeremy J. (1999), Geometry, Cambridge University Press, ISBN 978-0-521-59787-6

CHAPTER 8

Four-Dimensional Space Geometry

Chapter Contents

While giving someone a definition of the fourth dimension is relatively easy, giving someone an intuitive understanding of the fourth dimension can be quite difficult. A definition of the fourth dimension could go like this: The fourth dimension is all space that one can get to by travelling in a direction perpendicular to three-dimensional space. Whenever an uninitiated person hears this, they start pointing their finger around in the air, trying to figure out how it's possible for such a direction to exist. Such a short explanation gives them no intuitive feeling of the fourth dimension.

INTRODUCTION

In mathematics, four-dimensional space ("4D") is a geometric space with four dimensions. It typically is more specifically four-dimensional Euclidean space, generalizing the rules of three-dimensional Euclidean space. It has been studied by mathematicians and philosophers for over two centuries, both for its own interest and for the insights it offered into mathematics and related fields.

Algebraically, it is generated by applying the rules of vectors and coordinate geometry to a space with four dimensions. In particular, a vector with four components (a 4D vector or 4-tuple) can be used to represent a position in four-dimensional space. The space is a Euclidean space, so has a metric and norm, and so all directions are treated as the same: the additional dimension is indistinguishable from the other three.

In modern physics, space and time are unified in a four-dimensional Minkowski continuum called spacetime, whose metric treats the time dimension differently from the three spatial dimensions (see below for the definition of the Minkowski metric/pairing). Spacetime is not a Euclidean space.

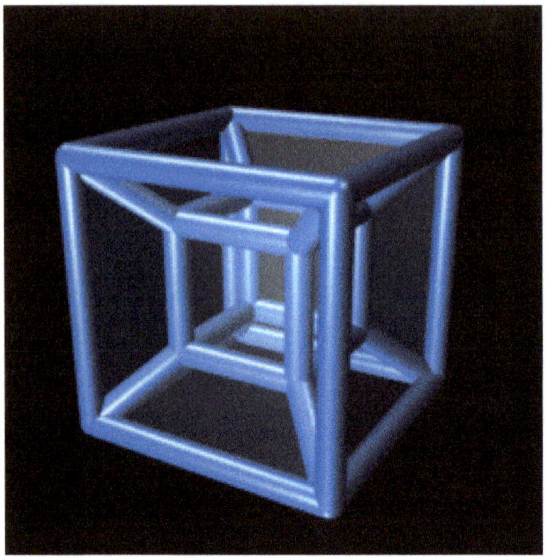

3D projection of a tesseract undergoing a simple rotation in four dimensional space.

Lagrange wrote in his *Mécanique analytique* (published 1788, based on work done around 1755) that mechanics can be viewed as operating in a four-dimensional space — three dimensions of space, and one of time. In 1827 Möbius realized that a fourth dimension would allow a three-dimensional form to be rotated onto its mirror-image, and by 1853 Ludwig Schläfli had discovered many polytopes in higher dimensions, although his work was not published until after his death. Higher dimensions were soon put on firm footing by Bernhard Riemann's 1854 Habilitationsschrift, *Über die Hypothesen welche der Geometrie zu Grunde liegen*, in which he considered a "point" to be any sequence of coordinates $(x_1, ..., x_n)$. The possibility of geometry in higher dimensions, including four dimensions in particular, was thus established.

An arithmetic of four dimensions called quaternions was defined by William Rowan Hamilton in 1843. This associative algebra was the source of the science of vector analysis in three dimensions as recounted in *A History of Vector Analysis*. Soon after tessarines and coquaternions were introduced as other four-dimensional algebras over **R**.

One of the first major expositors of the fourth dimension was Charles Howard Hinton, starting in 1880 with his essay *What is the Fourth Dimension?*; published in the Dublin

University magazine. He coined the terms *tesseract*, *ana* and *kata* in his book *A New Era of Thought*, and introduced a method for visualising the fourth dimension using cubes in the book *Fourth Dimension*.

Hinton's ideas inspired a fantasy about a "Church of the Fourth Dimension" featured by Martin Gardner in his January 1962 "Mathematical Games column" in *Scientific American*. In 1886 Victor Schlegel described his method of visualizing four-dimensional objects with Schlegel diagrams.

In 1908, Hermann Minkowski presented a paper consolidating the role of time as the fourth dimension of spacetime, the basis for Einstein›s theories of special and general relativity. But the geometry of spacetime, being non-Euclidean, is profoundly different from that popularised by Hinton. The study of Minkowski space required new mathematics quite different from that of four-dimensional Euclidean space, and so developed along quite different lines. This separation was less clear in the popular imagination, with works of fiction and philosophy blurring the distinction, so in 1973 H. S. M. Coxeter felt compelled to write:

Little, if anything, is gained by representing the fourth Euclidean dimension as *time*. In fact, this idea, so attractively developed by H. G. Wells in *The Time Machine*, has led such authors as John William Dunne (*An Experiment with Time*) into a serious misconception of the theory of Relativity. Minkowski's geometry of space-time is *not* Euclidean, and consequently has no connection with the present investigation.

—*H. S. M. Coxeter, Regular Polytopes*

Vectors

Mathematically four-dimensional space is simply a space with four spatial dimensions, that is a space that needs four parameters to specify a point in it. For example, a general point might have position vector **a**, equal to

$$\mathbf{a} = \begin{pmatrix} a_1 \\ a_2 \\ a_3 \\ a_4 \end{pmatrix}.$$

This can be written in terms of the four standard basis vectors ($\mathbf{e}_1$, $\mathbf{e}_2$, $\mathbf{e}_3$, $\mathbf{e}_4$), given by

$$\mathbf{e}_1 = \begin{pmatrix} 1 \\ 0 \\ 0 \\ 0 \end{pmatrix}; \mathbf{e}_2 = \begin{pmatrix} 0 \\ 1 \\ 0 \\ 0 \end{pmatrix}; \mathbf{e}_3 = \begin{pmatrix} 0 \\ 0 \\ 1 \\ 0 \end{pmatrix}; \mathbf{e}_4 = \begin{pmatrix} 0 \\ 0 \\ 0 \\ 1 \end{pmatrix},$$

so the general vector **a** is

$$\mathbf{a} = a_1\mathbf{e}_1 + a_2\mathbf{e}_2 + a_3\mathbf{e}_3 + a_4\mathbf{e}_4.$$

Vectors add, subtract and scale as in three dimensions.

The dot product of Euclidean three-dimensional space generalizes to four dimensions as

$$\mathbf{a} \cdot \mathbf{b} = a_1b_1 + a_2b_2 + a_3b_3 + a_4b_4.$$

It can be used to calculate the norm or length of a vector,

$$|\mathbf{a}| = \sqrt{\mathbf{a} \cdot \mathbf{a}} = \sqrt{{a_1}^2 + {a_2}^2 + {a_3}^2 + {a_4}^2},$$

and calculate or define the angle between two vectors as

$$\theta = \arccos \frac{\mathbf{a} \cdot \mathbf{b}}{|\mathbf{a}|\,|\mathbf{b}|}.$$

Minkowski spacetime is four-dimensional space with geometry defined by a nondegenerate pairing different from the dot product:

$$\mathbf{a} \cdot \mathbf{b} = a_1b_1 + a_2b_2 + a_3b_3 - a_4b_4.$$

As an example, the distance squared between the points (0,0,0,0) and (1,1,1,0) is 3 in both the Euclidean and Minkowskian 4-spaces, while the distance squared between (0,0,0,0) and (1,1,1,1) is 4 in Euclidean space and 2 in Minkowski space; increasing b_4 actually decreases the metric distance. This leads to many of the well known apparent "paradoxes" of relativity.

The cross product is not defined in four dimensions. Instead the exterior product is used for some applications, and is defined as follows:

$$\mathbf{a} \wedge \mathbf{b} = (a_1b_2 - a_2b_1)\mathbf{e}_{12} + (a_1b_3 - a_3b_1)\mathbf{e}_{13} + (a_1b_4 - a_4b_1)\mathbf{e}_{14} + (a_2b_3 - a_3b_2)\mathbf{e}_{23} + (a_2b_4 - a_4b_2)\mathbf{e}_{24} + (a_3b_4 - a_4b_3)\mathbf{e}_{34}.$$

This is bivector valued, with bivectors in four dimensions forming a six-dimensional linear space with basis ($\mathbf{e}_{12}$, $\mathbf{e}_{13}$, $\mathbf{e}_{14}$, $\mathbf{e}_{23}$, $\mathbf{e}_{24}$, $\mathbf{e}_{34}$). They can be used to generate rotations in four dimensions.

Orthogonality and Vocabulary

In the familiar 3-dimensional space in which we live there are three coordinate axes — usually labeled *x*, *y*, and *z* — with each axis orthogonal (i.e. perpendicular) to the other two. The six cardinal directions in this space can be called *up*, *down*, *east*, *west*, *north*, and *south*. Positions along these axes can be called *altitude*, *longitude*, and *latitude*. Lengths measured along these axes can be called *height*, *width*, and *depth*.

Comparatively, 4-dimensional space has an extra coordinate axis, orthogonal to the other three, which is usually labeled *w*. To describe the two additional cardinal directions, Charles Howard Hinton coined the terms *ana* and *kata*, from the Greek words meaning "up toward" and "down from", respectively. A position along the *w* axis can be called *spissitude*, as coined by Henry More.

Geometry

The geometry of 4-dimensional space is much more complex than that of 3-dimensional space, due to the extra degree of freedom.

Just as in 3 dimensions there are polyhedra made of two dimensional polygons, in 4 dimensions there are 4-polytopes made of polyhedra. In 3 dimensions there are 5 regular polyhedra known as the Platonic solids. In 4 dimensions there are 6 convex regular 4-polytopes, the analogues of the Platonic solids. Relaxing the conditions for regularity generates a further 58 convex uniform 4-polytopes, analogous to the 13 semi-regular Archimedean solids in three dimensions. Relaxing the conditions for convexity generates a further 10 nonconvex regular 4-polytopes.

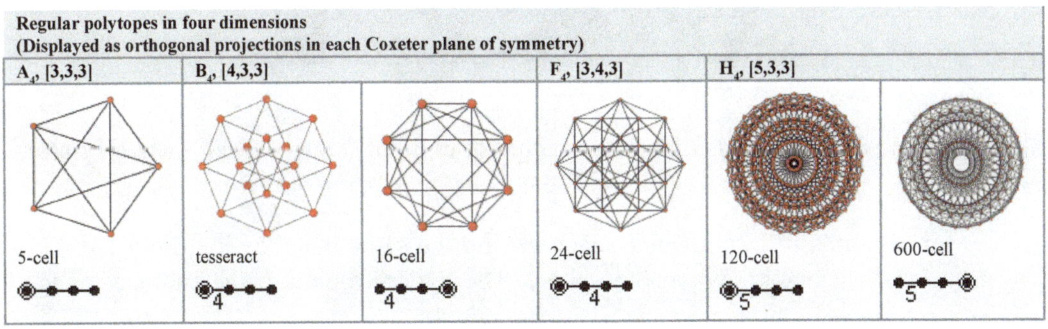

In 3 dimensions, a circle may be extruded to form a cylinder. In 4 dimensions, there are several different cylinder-like objects. A sphere may be extruded to obtain a spherical cylinder (a cylinder with spherical "caps", known as a spherinder), and a cylinder may be extruded to obtain a cylindrical prism (a cubinder). The Cartesian product of two circles may be taken to obtain a duocylinder. All three can "roll" in 4-dimensional space, each with its own properties.

In 3 dimensions, curves can form knots but surfaces cannot (unless they are self-intersecting). In 4 dimensions, however, knots made using curves can be trivially untied by displacing them in the fourth direction, but 2-dimensional surfaces can form non-trivial, non-self-intersecting knots in 4-dimensional space.Because these surfaces are 2-dimensional, they can form much more complex knots than strings in 3-dimensional space can. The Klein bottle is an example of such a knotted surface . Another such surface is the real projective plane.

Hypersphere

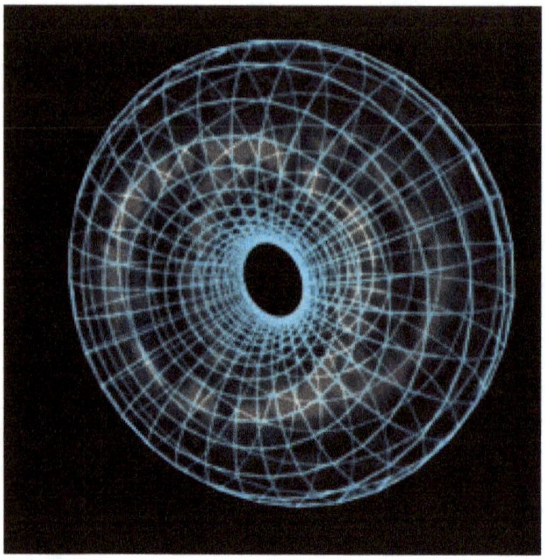

Stereographic projection of a Clifford torus: the set of points (cos(*a*), sin(*a*), cos(*b*), sin(*b*)), which is a subset of the 3-sphere.

The set of points in Euclidean 4-space having the same distance R from a fixed point P_0 forms a hypersurface known as a 3-sphere. The hyper-volume of the enclosed space is:

$$\mathbf{V} = \frac{1}{2}\pi^2 R^4$$

This is part of the Friedmann–Lemaître–Robertson–Walker metric in General relativity where R is substituted by function $R(t)$ with t meaning the cosmological age of the universe. Growing or shrinking R with time means expanding or collapsing universe, depending on the mass density inside.

Cognition

Research using virtual reality finds that humans, in spite of living in a three-dimensional world, can without special practice make spatial judgments based on the length of, and angle between, line segments embedded in four-dimensional space. The researchers noted that "the participants in our study had minimal practice in these tasks, and it remains an open question whether it is possible to obtain more sustainable, definitive, and richer 4D representations with increased perceptual experience in 4D virtual environments." In another study, the ability of humans to orient themselves in 2D, 3D and 4D mazes has been tested. Each maze consisted of four path segments of random length and connected with orthogonal random bends, but without branches or loops (i.e. actually labyrinths). The graphical interface was based on John McIntosh's free 4D Maze game. The participating persons had to navigate through the path and finally estimate the linear direction back to the starting point. The researchers found that some of the participants were able to mentally integrate their path after some practice in 4D (the lower-dimensional cases were for comparison and for the participants to learn the method).

Dimensional Analogy

A net of a tesseract

To understand the nature of four-dimensional space, a device called *dimensional analogy* is commonly employed. Dimensional analogy is the study of how $(n - 1)$ dimensions relate to n dimensions, and then inferring how n dimensions would relate to $(n + 1)$ dimensions.

Dimensional analogy was used by Edwin Abbott Abbott in the book *Flatland*, which narrates a story about a square that lives in a two-dimensional world, like the surface of a piece of paper. From the perspective of this square, a three-dimensional being has seemingly god-like powers, such as ability to remove objects from a safe without breaking it open (by moving them across the third dimension), to see everything that from the two-dimensional perspective is enclosed behind walls, and to remain completely invisible by standing a few inches away in the third dimension.

By applying dimensional analogy, one can infer that a four-dimensional being would be capable of similar feats from our three-dimensional perspective. Rudy Rucker illustrates this in his novel *Spaceland*, in which the protagonist encounters four-dimensional beings who demonstrate such powers.

Cross-Sections

As a three-dimensional object passes through a two-dimensional plane, a two-dimensional being would only see a cross-section of the three-dimensional object. For example, if a spherical balloon passed through a sheet of paper, a being on the paper would see first a single point, then a circle gradually growing larger, then smaller again until it shrank to a point and then disappeared. Similarly, if a four-dimensional object passed through three dimensions, we would see a three-dimensional cross-section of the four-dimensional object—for example, a hypersphere would appear first as a point, then as a growing sphere, with the sphere then shrinking to a single point and then disappearing. This means of visualizing aspects of the fourth dimension was used in the novel *Flatland* and also in several works of Charles Howard Hinton.

Projections

A useful application of dimensional analogy in visualizing the fourth dimension is in projection. A projection is a way for representing an n-dimensional object in $n - 1$ dimensions. For instance, computer screens are two-dimensional, and all the photographs of three-dimensional people, places and things are represented in two dimensions by projecting the objects onto a flat surface. When this is done, depth is removed and replaced with indirect information. The retina of the eye is also a two-dimensional array of receptors but the brain is able to perceive the nature of three-dimensional objects by inference from indirect information (such as shading, foreshortening, binocular vision, etc.). Artists often use perspective to give an illusion of three-dimensional depth to two-dimensional pictures.

Similarly, objects in the fourth dimension can be mathematically projected to the familiar 3 dimensions, where they can be more conveniently examined. In this case, the ‹retina› of the four-dimensional eye is a three-dimensional array of receptors. A hypothetical being with such an eye would perceive the nature of four-dimensional objects by inferring four-dimensional depth from indirect information in the three-dimensional images in its retina.

The perspective projection of three-dimensional objects into the retina of the eye introduces artifacts such as foreshortening, which the brain interprets as depth in the third dimension. In the same way, perspective projection from four dimensions produces similar foreshortening effects. By applying dimensional analogy, one may infer four-dimensional depth from these effects.

As an illustration of this principle, the following sequence of images compares various views of the 3-dimensional cube with analogous projections of the 4-dimensional tesseract into three-dimensional space.

Cube	Tesseract	Description
		The image on the left is a cube viewed face-on. The analogous viewpoint of the tesseract in 4 dimensions is the **cell-first perspective projection**, shown on the right. One may draw an analogy between the two: just as the cube projects to a square, the tesseract projects to a cube. Note that the other 5 faces of the cube are not seen here. They are *obscured* by the visible face. Similarly, the other 7 cells of the tesseract are not seen here because they are obscured by the visible cell.
	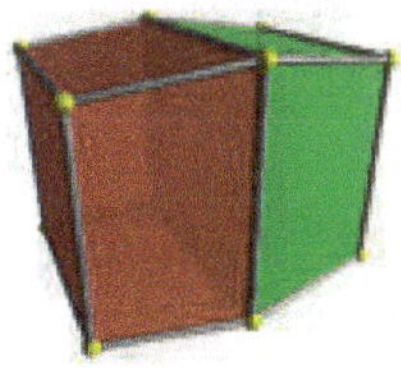	The image on the left shows the same cube viewed edge-on. The analogous viewpoint of a tesseract is the **face-first perspective projection**, shown on the right. Just as the edge-first projection of the cube consists of two trapezoids, the face-first projection of the tesseract consists of two frustums. The nearest edge of the cube in this viewpoint is the one lying between the red and green faces. Likewise, the nearest face of the tesseract is the one lying between the red and green cells.
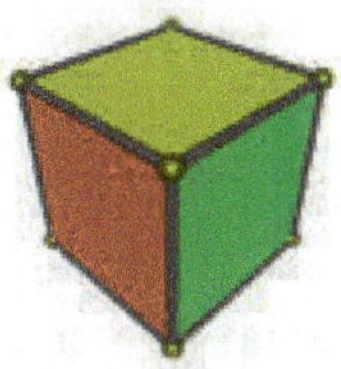	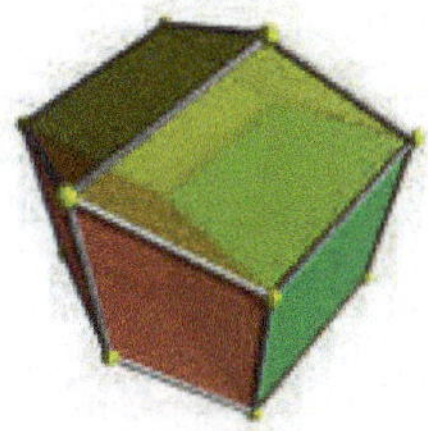	On the left is the cube viewed corner-first. This is analogous to the **edge-first perspective projection** of the tesseract, shown on the right. Just as the cube's vertex-first projection consists of 3 deltoids surrounding a vertex, the tesseract's edge-first projection consists of 3 hexahedral volumes surrounding an edge. Just as the nearest vertex of the cube is the one where the three faces meet, so the nearest edge of the tesseract is the one in the center of the projection volume, where the three cells meet.

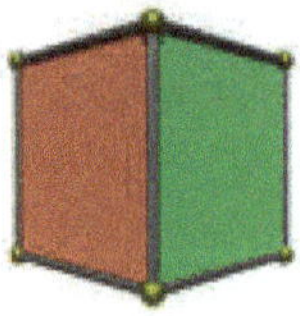	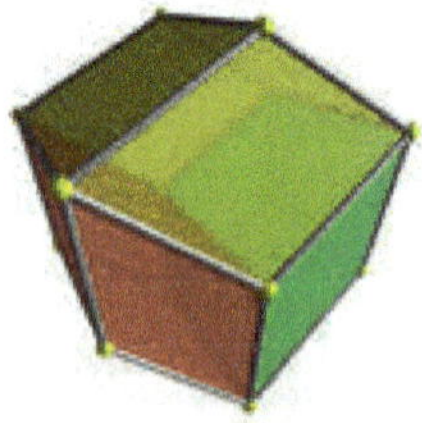	A different analogy may be drawn between the edge-first projection of the tesseract and the edge-first projection of the cube. The cube's edge-first projection has two trapezoids surrounding an edge, while the tesseract has *three* hexahedral volumes surrounding an edge.
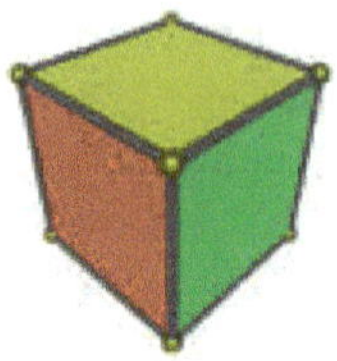	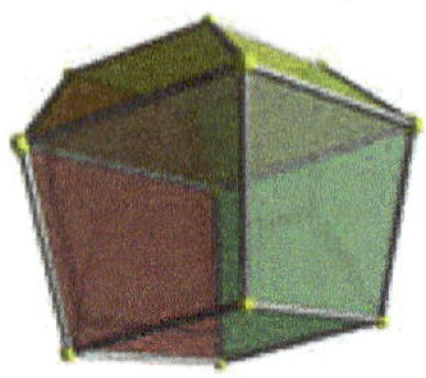	On the left is the cube viewed corner-first. The **vertex-first perspective projection** of the tesseract is shown on the right. The cube's vertex-first projection has three tetragons surrounding a vertex, but the tesseract's vertex-first projection has *four* hexahedral volumes surrounding a vertex. Just as the nearest corner of the cube is the one lying at the center of the image, so the nearest vertex of the tesseract lies not on boundary of the projected volume, but at its center *inside*, where all four cells meet. Note that only three faces of the cube's 6 faces can be seen here, because the other 3 lie *behind* these three faces, on the opposite side of the cube. Similarly, only 4 of the tesseract's 8 cells can be seen here; the remaining 4 lie *behind* these 4 in the fourth direction, on the far side of the tesseract.

Shadows

A concept closely related to projection is the casting of shadows.

If a light is shone on a three-dimensional object, a two-dimensional shadow is cast. By dimensional analogy, light shone on a two-dimensional object in a two-dimensional world would cast a one-dimensional shadow, and light on a one-dimensional object in a one-dimensional world would cast a zero-dimensional shadow, that is, a point of non-light. Going the other way, one may infer that light shone on a four-dimensional object in a four-dimensional world would cast a three-dimensional shadow.

If the wireframe of a cube is lit from above, the resulting shadow is a square within a square with the corresponding corners connected. Similarly, if the wireframe of a tesseract were lit from "above" (in the fourth dimension), its shadow would be that of a three-dimensional cube within another three-dimensional cube. (Note that, technically, the visual representation shown here is actually a two-dimensional image of the three-dimensional shadow of the four-dimensional wireframe figure.)

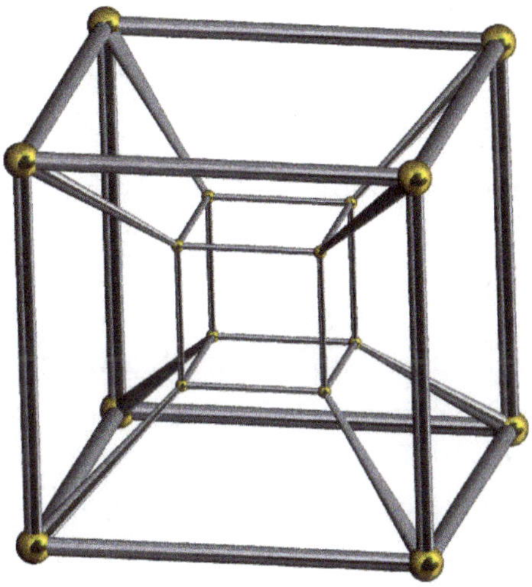

Bounding Volumes

Dimensional analogy also helps in inferring basic properties of objects in higher dimensions. For example, two-dimensional objects are bounded by one-dimensional boundaries: a square is bounded by four edges. Three-dimensional objects are bounded by two-dimensional surfaces: a cube is bounded by 6 square faces. By applying dimensional analogy, one may infer that a four-dimensional cube, known as a tesseract, is bounded by three-dimensional volumes. And indeed, this is the case: mathematics shows that the tesseract is bounded by 8 cubes. Knowing this is key to understanding how to interpret a three-dimensional projection of the tesseract. The boundaries of the tesseract project to *volumes* in the image, not merely two-dimensional surfaces.

Visual Scope

Being three-dimensional, we are only able to see the world with our eyes in two dimensions. A four-dimensional being would be able to see the world in three dimensions. For example, it would be able to see all six sides of an opaque box simultaneously, and in fact, what is inside the box at the same time, just as we can see the interior of a square on a piece of paper. It would be able to see all points in 3-dimensional space simultaneously, including the inner structure of solid objects and things obscured from our three-dimensional viewpoint. Our brains receive images in the second dimension and use reasoning to help us "picture" three-dimensional objects.

Limitations

Reasoning by analogy from familiar lower dimensions can be an excellent intuitive guide, but care must be exercised not to accept results that are not more rigorously tested. For example, consider the formulas for the circumference of a circle $C = 2\pi r$ and the surface area of a sphere: $A = 4\pi r^2$. One might be tempted to suppose that the surface volume of a hypersphere is $V = 6\pi r^3$, or perhaps $V = 8\pi r^3$, but either of these would be wrong. The correct formula is $V = 2\pi^2 r^3$.

REFERENCES

1. Bell, E.T. (1937). Men of Mathematics, Simon and Schuster, p. 154.
2. Coxeter, H.S.M. (1973). Regular Polytopes (3rd ed.). New York: Dover Publishing. ISBN 0486614808.
3. Rucker, Rudolf v.B. (1980). Speculations on the Fourth Dimension: Selected writings of Charles H. Hinton. New York: Dover. pp. vii. ISBN 0-486-23916-0.
4. Hinton, Charles Howard (1996). The Forth Dmension (1904). Kessinger Publising. ISBN 1-5645-9708-3.
5. Gardner, Martin (1975). Mathematical Carnival: From Penny Puzzles. Card Shuffles and Tricks of Lightning Calculators to Roller Coaster Rides into the Fourth Dimension (1st ed.). New York: Knopf. pp. 42, 52–53. ISBN 0-394-49406-7.
6. Victor Schlegel (1886) Ueber Projectionsmodelle der regelmässigen vier-dimensionalen Körper, Waren
7. Minkowski, Hermann (1909), "Raum und Zeit", Physikalische Zeitschrift, 10: 75–88
8. Various English translations on Wikisource: Space and Time
9. Møller, C. (1972). The Theory of Relativity (2nd ed.). Oxford: Clarendon Press. p. 93. ISBN 0-19-851256-2.
10. Carter, J.Scott; Saito, Masahico. Knotted Surfaces and Their Diagrams. American Mathematical Society. ISBN 0821874918.
11. D'Inverno, Ray (1998). Introducing Einstein's Relativity (Reprint ed.). Oxford: Clarendon Press. p. 319. ISBN 0-19-859653-7.
12. Ambinder, Michael S.; Wang, Ranxiao Frances; Crowell, James A.; Francis, George K.; Brinkmann, Peter (October 2009). "Human four-dimensional spatial intuition in virtual reality". Psychonomic Bulletin & Review. 16 (5): 818–823. doi:10.3758/PBR.16.5.818. PMID 19815783.
13. Aflalo, T. N.; Graziano, M. S. A. (2008). "Four-dimensional spatial reasoning in humans." (PDF). Journal of Experimental Psychology: Human Perception and Performance. 34 (5): 1066–1077. doi:10.1037/0096-1523.34.5.1066.
14. "4D Maze Game". urticator.net. Retrieved 2016-12-16.

15. Kaku, Michio (1995). Hyperspace: A Scientific Odyssey Through Parallel Universes, Time Warps, and the Tenth Dimension (reissued ed.). Oxford: Oxford University Press. pp. Part I, Chapter 3. ISBN 0-19-286189-1.

16. Rucker, Rudy (1996). The Fourth Dimension: A Guided Tour of the Higher Universe. Boston: Houghton Mifflin. p. 18. ISBN 0-395-39388-4.

Index